For all single parents who put as much energy and love into bringing up their children as our mother did for my sister and me.

And for Hedi.

10th
ANNIVERSARY EDITION

GUT

The Book That Began
the Microbiome
Revolution

Giulia Enders
Translated by David Shaw

Melbourne | London | Minneapolis

Scribe Publications
18–20 Edward St, Brunswick, Victoria 3056, Australia
2 John St, Clerkenwell, London, WC1N 2ES, United Kingdom
3754 Pleasant Ave, Suite 223w, Minneapolis, Minnesota 55409, USA

Originally published © 2014 by Ullstein Buchverlage GmbH, Berlin
First published in English by Scribe 2015
Revised edition published 2017
This edition published 2025

Text and Illustrations copyright © by Ullstein Buchverlage GmbH, Berlin.
Published in 2017 by Ullstein Verlag

Translation copyright © David Shaw 2014, 2017

All rights reserved. Without limiting the rights under copyright reserved above, no part of this publication may be reproduced, stored in or introduced into a retrieval system, or transmitted, in any form or by any means (electronic, mechanical, photocopying, recording or otherwise) without the prior written permission of the publishers of this book.

The advice provided in this book has been carefully considered and checked by the author and the publisher. It should not, however, be regarded as a substitute for competent medical advice. Therefore, all information in this book is provided without any warranty or guarantee on the part of the publisher or the author. Neither the author nor the publisher or their representatives shall bear any liability whatsoever for personal injury, property damage, and financial losses.

The moral rights of the author and translator have been asserted.

Typeset in Kepler Light by the publishers

Printed and bound in the UK by CPI Group (UK) Ltd, Croydon CR0 4YY

Scribe is committed to the sustainable use of natural resources and the use of paper products made responsibly from those resources.

978 1 925106 67 1 (AU paperback)
978 1 917189 31 6 (UK paperback)
978 1 925113 78 5 (ebook)

Catalogue records for this book are available from the National Library of Australia and the British Library.

scribepublications.com.au
scribepublications.co.uk
scribepublications.com

Contents

Foreword ... 1

1 Gut Feeling ... 7

How does pooing work? ... and why that's an important question ... 10
    Are you sitting properly? ... 14

The Gateway to the Gut ... 19

The Structure of the Gut ... 29
    The 'gargly' oesophagus ... 30
    The lopsided stomach pouch ... 32
    The meandering small intestine ... 33
    The unnecessary appendix and
        the bulgy large intestine ... 39

What We Really Eat ... 44

Allergies and Intolerances ... 53
    Coeliac disease and gluten sensitivity ... 53
    Lactose intolerance
        and fructose intolerance ... 56

A Few Facts about Faeces ... 63

2 The Nervous System of the Gut — 73

How Our Organs Transport Food	75
Eyes	75
Nose	75
Mouth	76
Pharynx	77
Oesophagus	77
Stomach	80
Small intestine	81
Large intestine	84
Reflux	87
Vomiting	92
Why we vomit and what we can do to prevent it	93
Constipation	101
Laxatives	107
The three-day rule	113
The Brain and the Gut	114
How the gut influences the brain	117
Of irritated bowels, stress, and depression	121
Where the 'self' originates	131

3 The World of Microbes — 135

I Am an Ecosystem	137

The Immune System and Our Bacteria	141
The Development of the Gut Flora	147
The Adult Gut Population	155
The genes of our bacteria	160
The three gut types	162
The Role of the Gut Flora	169
How might bacteria make us fat?	
Three theories	174
Cholesterol and gut bacteria	178
The Bad Guys: harmful bacteria and parasites	183
Salmonellae in hats	183
Helicobacter: humanity's first 'pet'	188
Toxoplasmata: fearless cat riders	196
Worms	204
Of Cleanliness and Good Bacteria	209
Everyday cleanliness	210
Antibiotics	218
Probiotics	225
Prebiotics	237
The Gut-Brain Axis	247
New discoveries	248
Clever cravings for fermented foods	259
Acknowledgements	264
Main References	265

Foreword

I was born by caesarean section, and was not able to be breast-fed. That makes me a perfect poster child for the intractability of the gastro-intestinal tract in the 21st century. If I had known more about the gut back then, I could have placed bets on what illnesses I would contract in later life. At first I was lactose intolerant. I never thought about why I was suddenly able to drink milk again at the age of five; at some point I got fat, then thin again. Then, for a long time, I was fine. Until I got 'the sore'.

When I was seventeen, I developed a small sore on my right leg, for no apparent reason. It stubbornly refused to heal, and after a month I went to see my doctor. She didn't really know what it was, and prescribed me some cream. Three weeks later, my entire leg was covered in sores. Soon they spread to my other leg, my arms, and my back. Sometimes they appeared on my face. Luckily, it was winter at the time, and everyone thought I had cold sores and a graze on my forehead.

No doctor was able to help me — giving me vague diagnoses of some kind of nervous eczema. They asked me about stress and psychological problems. Cortisone helped a little, but as soon as I stopped using it, the sores just came back. For a whole year, in summer and in winter, I wore tights to stop my sores from weeping though my trousers. Then I pulled myself together and started doing some research of my own. By chance, I came across a report about a very similar skin condition. A man had contracted it after taking antibiotics, and I, too, had had to take a course of antibiotics just a couple of weeks before my first sore appeared.

From that moment on, I ceased to treat my skin like that of a person with a dermatological problem, and began to see it as the skin of a person with an intestinal condition. I stopped eating dairy products, cut out gluten almost entirely, swallowed various bacteria cultures, and generally improved my diet. I also carried out some pretty crazy experiments on myself ... if I had already been studying medicine at that time, I wouldn't have dared do half of them. Once, I overdosed on zinc for several weeks, causing me to have an extremely heightened sense of smell for the next few months.

With a few tricks, I finally managed to get my condition under control. This success gave me a lift, and I experienced with my own body the reality that knowledge is power. That's when I started studying medicine.

In my first semester as a student, I was at a party where I ended up sitting next to a guy who had the smelliest breath I have ever smelled. It wasn't a typical bad-breath smell — not the scratchy hydrogen-breath odours of stressed-out middle-aged gentlemen, nor the sugary-foetid funk from the mouth of an elderly aunt with too sweet a tooth. After a while, I moved away and sat somewhere else. The next day, he was dead. He had killed himself. I couldn't get him out of my mind. Could it have been a diseased gut creating that smell, and if so, could a diseased gut also have affected his psychological state?

A week later, I decided to share my suspicion with a good friend. And a few months after that, the same friend contracted a bad case of gastroenteritis, which left her feeling very poorly. The next time we met, she told me she thought there might be something in my theory, as her illness had made her feel worse than she ever had before, psychologically as well as physically. Her comments inspired me to start looking more closely at this subject matter. Soon, I discovered there was an entire branch of medical research investigating the links between the gut and the brain. It's a rapidly growing field of study. Ten or so years ago, there were hardly any published studies on the subject, but now there are several hundred academic articles covering the field. The influence of the gut on our health and wellbeing is one of the *new* lines of research in modern medicine! The renowned American biochemist Rob Knight told the journal *Nature* that the field offered at least as much promise as stem-cell research. I had stumbled upon a subject I found more and more fascinating.

As I continued my medical degree, I realised how neglected, even looked down upon, this area is in the medical world. This is all the more surprising when you consider what an extraordinary organ the gut is. It accounts for two-thirds of our immune system, extracts energy from sandwiches and vegetarian sausages, and produces more than twenty unique hormones. Most doctors learn very little about this in their training. When I visited the 'Microbiome and Host Health' symposium in Lisbon in May 2013, the number of participants was modest. About half came from institutions with the financial wherewithal to allow them to be among the 'pioneers', including Harvard, Yale, Oxford, and EMBL Heidelberg.

I'm sometimes shocked by the way scientists huddle behind closed doors to discuss their important research results, without informing the public about them at all. Academic caution is often preferable to premature publication. But fear can also destroy opportunities. It is now generally accepted in scientific circles that people with certain digestive problems often suffer from nervous disorders of the gut. Their gut then sends signals to the part of the brain that processes negative feelings, although they have done nothing bad. Such patients feel uneasy, but have no idea why. If their doctors simply treat them as irrational mental cases, it can be extremely counterproductive. And this is just one example of why some research results should be published more quickly.

And that is my aim in writing this book: I want to make new

knowledge available to a broad audience, and communicate the information that scientists bury in their academic publications or discuss behind closed doors at scientific meetings, while many ordinary people out there are searching for answers. I know there are many patients suffering from unpleasant conditions who are frustrated by the medical world. I can't offer any panaceas, and keeping your gut healthy is not a miracle cure for everything. But what I can do is to show, in an entertaining way, why the gut is so fascinating, what exciting new research is currently underway and how we can use this new knowledge to improve our daily lives.

My medical studies and my doctoral research at the Institute for Medical Microbiology have given me the skills to sift and sort scientific data. My own personal experience has helped me develop the ability to communicate this knowledge to people. My sister has given me the support I needed to keep me on the right track — listening to me read aloud from my manuscript and saying, with a charming grin, 'I think you better try that bit again.'

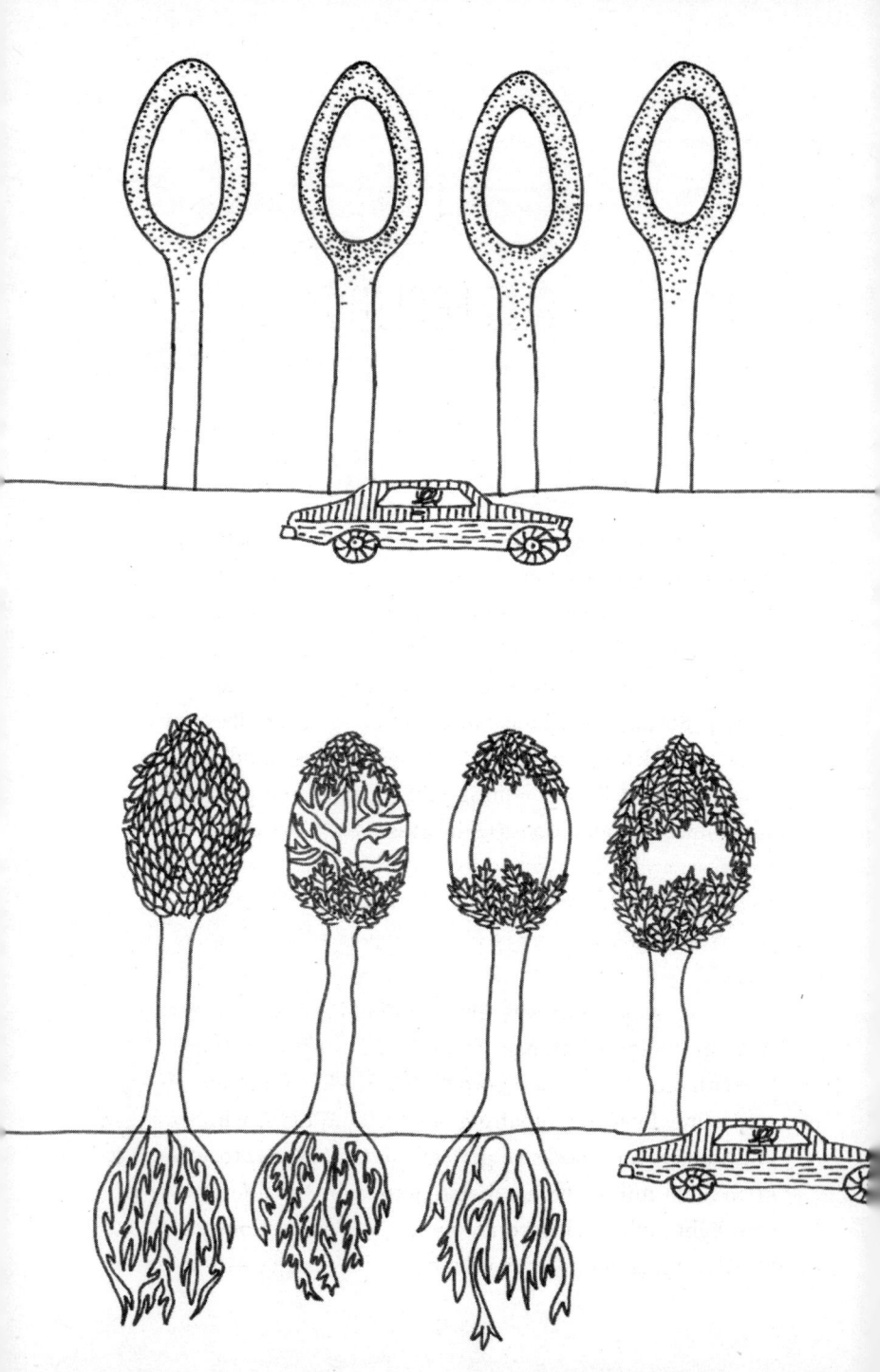

1
Gut Feeling

The world is a much more interesting place if we look beyond what is visible to the naked eye — there is so much more to see! If we start to look more closely, a tree can be more than a spoon-shaped thing. In a highly simplified way, 'spoon' is the general shape we perceive when we look at a tree: a straight trunk and a round treetop. Seeing that shape, our eyes tell us 'spoon-like thing'. But there are at least as many roots beneath the ground as there are branches above it. Our brain should really be telling us something like 'dumbbell', but it doesn't. The brain gets most of its input from our eyes, and that information is very rarely in the form of an illustration in a book showing trees in their entirety. So it faithfully construes a passing forest landscape as 'spoon, spoon, spoon, spoon'.

As we 'spoon' our way through life like this, we overlook all sorts of wonderful things. There is a buzz of constant activity beneath our skin. We are perpetually flowing, pumping, sucking, squeezing, bursting, repairing, and rebuilding. A whole crew of ingenious organs works so perfectly and efficiently together that, in an adult human being, they require as much energy as a 100-watt light bulb. Each second, our kidneys meticulously filter our blood — much more efficiently than a coffee filter — and in most cases they carry on doing so for our entire lives. Our lungs are so cleverly designed that we use energy only when we breathe in. Breathing out happens without any expenditure of energy at all. If we were transparent, we would be able to see the beauty of this mechanism: like a wind-up toy car, only bigger, softer and more lung-y. While some of us might be sitting around thinking *Nobody cares about me!*, our heart is currently working its seventeen thousandth 24-hour shift — and would have every right to feel a little forgotten when its owner thinks such thoughts.

If we could see more than meets the eye, we could watch as a clump of cells grows into a human being in a woman's tummy. We would suddenly see how we develop, roughly speaking, from three 'tubes'. The first tube runs right the way through us, with a knot in the middle. This is our cardiovascular system, and the central knot is what develops into our heart. The second tube develops more or less parallel to the first along our back, then forms a bubble that migrates to the top end of our body, where it stays put. This tube is our nervous system, with the spinal cord including the brain at the top, and myriad nerves branching out into every part of our body. The third tube runs through us from end to end. This is our intestinal tube — the gut.

The intestinal tube provides many of the furnishings of our interior. It grows buds that bulge out farther and farther to the right and left. These buds will later develop into our lungs. A little bit lower down, the intestinal tube bulges again, and our liver has begun to develop. It also forms our gall bladder and pancreas. But, most importantly, the tube itself begins to grow increasingly clever. It is involved in the complex construction of our mouth, creates our oesophagus, with its ability to 'breakdance', and develops a little stomach pouch, so we can store food for a couple of hours. And, last but not least, the intestinal tube completes its masterpiece — the eponymous intestine, or gut.

The 'masterpieces' of the other two tubes — the heart and

the brain — are generally held in high regard. We see the heart as central to life since it pumps blood around the body; the brain is admired for its ability to create a firework of new mental images and concepts every second. But the gut, in most people's eyes, is good for little more than going to the loo. Apart from that, people think, it just hangs around inside our bellies, letting off a little 'steam' every now and then. People do not generally ascribe any particular abilities to it. It would be fair to say that we underestimate our gut — or, to put it more bluntly, we don't just underestimate it, we are ashamed of it: more 'guilt feeling' than 'gut feeling'!

I hope this book will change that — by making use of the wonderful ability that books possess to show us more than the world we see around us: trees are not spoons! And a gut feeling is a good feeling!

How does pooing work?
... and why that's an important question

My flatmate wandered into the kitchen one day, saying, 'Giulia, you study medicine — so how does pooing work?' It probably wouldn't be a great idea to begin my autobiography with that question, but that little query did literally change my life. I withdrew to my room, sat on the floor, and was soon poring over three different textbooks. The answer I eventually discovered left me flabbergasted. This unspectacular daily necessity turned out to be far more sophisticated and impressive than I ever would have imagined.

Every time we go to the loo, it's a masterly performance — two nervous systems working tirelessly in tandem to dispose of our waste as discreetly and hygienically as possible. Very few other animals do their business in such an admirable and orderly manner. Our bodies have developed all sorts of mechanisms and techniques to help us poo properly. The first surprise is the sophistication of our sphincters. The vast majority of people are familiar only with the outer sphincter; the muscle we can consciously control, opening and closing it at will. There is another, very similar, muscle just centimetres away — but this is one we can't control consciously.

Each of the two sphincters looks after the interests of a different nervous system. The outer muscle is a faithful servant of our consciousness. When our brain deems it an unsuitable time to go to the toilet, the external sphincter obeys and stays closed with all its might. The internal sphincter represents our unconscious inner world. Whether Great Aunt Bertha approves of breaking wind or not is of no concern to the *sphincter ani internus*. It is only interested in making sure everything is okay inside us. Is the gas pressure rising? The inner sphincter's mission is to keep all unpleasantness at bay. If it had its way, Great Aunt Bertha would break wind more often. The main thing for the internal sphincter is to keep everything comfortable and in its place.

These two sphincter muscles have to work as a team. When what's left of our food reaches the internal sphincter, that muscle's reflex response is to open. But it does not just open the floodgates and let everything out, leaving the outer sphincter to deal with the onrush. First, it allows a small 'taster' through. The space between the internal and external sphincter muscles is home to a large number of sensor cells. They analyse the product delivered to them, test it to find out whether it is solid or gaseous, and send the resulting information up to the brain. This is the moment when the brain realises: it's time to go to the toilet! ... or maybe it's just a bit of wind? It then does what it is so good at, with its conscious awareness: it adapts to the environment we find ourselves in. It compares the information it receives from our eyes and ears to the data in its memory bank of past experiences. In this way, the brain takes just a matter of seconds to make an initial assessment of the situation and send a message back to the sphincter: 'I've had a look, and we're at Great Aunt Bertha's, in the living-room — we might get away with breaking a little wind, if we can squeeze it out silently. Anything more solid might not be such a good idea.'

The external sphincter gets the message, and dutifully squeezes itself closed even more tightly than before. The internal sphincter receives this signal from its more outgoing partner and respects the decision — for now. The two muscles work together and manoeuvre the 'taster' back into the holding pattern. Of course, it will have to come out sooner or later, just not here and not now. After a while, the internal sphincter will simply give it a try with another little 'taster'. If by then we're back within our familiar four walls, it's 'full steam ahead'!

Our internal sphincter is a no-nonsense little guy. His motto is, 'If it's gotta come out, it's gotta come out!' Not much room for argument there. The external sphincter, on the other hand, has to deal with the vagaries of the outside world and its many options: it might, theoretically, be possible to use this stranger's toilet, but is that a good idea? Have my new girlfriend/boyfriend and I been together long enough for farting in front of each other to be okay — and if so, is it down to me to break the ice and go first? If I don't go to the loo now, can I wait till this evening, or will I get caught short?

The considerations of our sphincter may not sound worthy of a Nobel Prize, but in fact they are concerned with some of the most basic questions of human existence: how important to us is our inner world, and what compromises should we make to get by in the external world? There are those who clench with all their might to keep the wind in, come what may, eventually struggling home wracked with belly ache. Others get granny to pull on their finger at a family party, making a funny, if slightly vulgar, magic show out of their need to break wind. In the long run, the best compromise is probably somewhere in the middle.

If we suppress our need to go the loo too often or for too long, our internal sphincter begins to feel browbeaten. In fact, we are able to re-educate it completely. That means the sphincter and the surrounding muscles have been disciplined so often by the external sphincter that they become cowed. If communication between the two sphincters breaks down completely, constipation can result.

Even without such defecatory discipline, something very similar can happen to women during labour. Childbirth can cause tearing of the delicate nerve fibres that allow the two muscles to communicate with each other. The good news is that those nerves can heal and reconnect. Irrespective of whether the damage was caused by childbirth or some other way, one good treatment option is what doctors call biofeedback therapy. It teaches the two sphincters to overcome their estrangement and get to know each other again. A machine is used to measure how efficiently the internal and external sphincter are working together. If messages from one to the other get through, the patient is rewarded with a sound or light signal. It's like one of those quiz shows on early-evening TV, where the whole set lights up and fanfares blare when a contestant gets the answer right — only it's at a medical practice, not on TV, and the 'contestant' has a sensor electrode up their bum. That may seem extreme, but it's worth it. When the two sphincters are talking properly to each other again, going to the little girls' or little boys' room is an altogether more pleasant experience.

Sphincters, sensor cells, consciousness, and electrode-up-the-bum quiz shows — my flatmate was probably not expecting all that in answer to his casual question about pooing. Nor did the group of rather prim female business-studies students who had meanwhile gathered in the kitchen for his birthday tea party. Still, the evening turned out to be fun, and it made me realise that a lot of people are actually interested in the gut. Some interesting new questions were raised at the birthday party: is it true that we don't sit on the toilet properly? How can we burp more easily? Why can we get energy from steaks, apples, or fried potatoes, for example, but a car can only run on one kind of fuel? Why do we have an appendix? Why are faeces always the same colour?

My flatmates have learned to recognise the familiar look on my face when I rush into the kitchen, bursting to tell them my latest gut anecdote — like the one about the tiny squat toilets and luminous stools.

Are you sitting properly ...?

It's a good idea to question your own habits from time to time. Are you really taking the shortest and most interesting route to the bus stop? Is that comb-over to hide your increasing bald patch effective and chic? Or, indeed, are you sitting properly when you go to the toilet?

There will not always be a clear, unambiguous answer to every question, but a little experimentation can sometimes open up whole new vistas. That is probably what was going through the mind of Dov Sikirov when the Israeli doctor asked 28 test subjects to do their daily business in three alternative positions: enthroned on a normal toilet; half-sitting, half-squatting on an unusually low toilet; and squatting with no seat beneath them at all. He recorded the time they took in each position, and asked the volunteers to assess the degree of straining that their bowel movements had required. The results were clear: in a squatting position, the subjects took an average of 50 seconds, and reported a feeling of full, satisfactory bowel emptying. The average time when seated was 130 seconds, and the resulting feeling was not deemed to be quite so satisfactory. (Anyway, tiny little toilets look kind of cute, whatever you do on them.)

Why is this? The closure mechanism of our gut is not designed in such a way that it can open the hatch completely when we are seated. There is a muscle that encircles the gut like a lasso when we are sitting or, indeed, standing, and pulls it in one direction, creating a kink in the tube. This mechanism is a kind of extra insurance policy, in addition to our old friends, the sphincters.

Some people will be familiar with this 'kinky' closing mechanism from their garden hose. You ask your sister to check why there's no water coming out of the hose. When she peers down the end, you quickly unbend the kink, and it's just a few minutes till your parents ground you for a week.

But back to our 'kinky' rectal-closure mechanism: it means our faeces hit a corner. Just like a car on the highway, turning a corner means it has to put on the brakes. So, when we are sitting or standing, our sphincters have to expend much less energy keeping everything in. If the lasso muscle relaxes, the kink straightens. The road ahead is straight, and the faeces are free to 'step on the gas'.

Squatting has been the natural pooing position for humans since time immemorial. The modern sitting toilet has existed only since indoor sanitation became common, in the late eighteenth century. But such 'cavemen did it that way' arguments are often met with distain by the medical profession. Who says that squatting helps the muscle relax better and straightens the faeces highway? Japanese researchers fed volunteers luminescent substances and X-rayed them while doing their business in various positions. They found out two interesting things. First, squatting does indeed lead to a nice, straight intestinal tract, allowing for a direct, easy exit. Second, some people are nice enough to let researchers feed them luminous substances and X-ray them while they poo, all in the name of science. Both findings are pretty impressive, I think.

Haemorrhoids, digestive diseases like diverticulitis, and even constipation are common only in countries where people generally sit on some kind of chair to pass their stool. This is not due to lack of tissue strength, especially in young people, but to the fact that there is too much pressure on the end of the gut. Some people tend to tense up their entire belly muscles when they are stressed. Often, they don't even realise they are doing it. Haemorrhoids prefer to avoid internal pressure like that, by dangling loosely out of the anus. Diverticula are small, light-bulb-shaped pouches in the bowel wall, resulting from the tissue in the gut bulging outwards under pressure.

Of course, the way we go to the toilet is not the only cause of haemorrhoids and diverticula. However, it remains a fact that the 1.2 billion people in this world who squat have almost no incidence of diverticulosis, and far fewer problems with haemorrhoids. We in the West, on the other hand, squeeze our gut tissue till it comes out of our bottoms and we have to have it removed by a doctor. Do we put ourselves through all this just because sitting on a throne is more 'civilised' than silly squatting? Doctors believe that straining too much or too often on the toilet can also seriously increase the risk of varicose veins, a stroke, or defecation syncope — fainting on the toilet.

A text message I received from a friend who was on holiday in France read, 'The French are crazy! Someone's stolen the toilets from the last three service stations we stopped at!' I had to laugh: first, because I suspected my friend was actually being serious, and second, because it reminded me of my first experience of French squat toilets. *Why am I being forced to squat here when you could just as easily have put in a proper toilet?* I mournfully complained to myself as I recovered from the shock of the emptiness I saw before me. Throughout much of Asia, Africa, and southern Europe, people squat briefly over such toilets in a kind of martial arts or downhill skiing pose to poo. We, by contrast, take so long, we have to while away the time till we've finished our business with reading the paper, carefully pre-folding pieces of toilet paper for imminent use, scanning the corners of the bathroom to see if they could do with a clean, or staring patiently at the opposite wall.

When I read this chapter out to my family in our living room, I looked up to see disconcerted faces. Are we going to have to descend from our porcelain thrones and squat precariously over a hole to poo? Of course not, haemorrhoids or no haemorrhoids! Although it might be fun to try climbing up onto the toilet seat

to do our business while squatting there. But there's no need for that, either — it is possible to squat while sitting. It's a particularly good idea when things don't come so easy, so to speak. To do it, just incline your upper body forward slightly and place your feet on a low footrest placed in front of the toilet, and — *voilà* — all the angles are correct, and you can read the paper, pre-fold your tissue, or stare at the wall with a clear conscience.

The Gateway to the Gut

You might think that the back end of the gut holds so many surprises in store for us because it is something we do not think about very much. But I don't think that's the real reason. The other end of the gut, the gateway, so to speak, also has no shortage of surprises in store — although we are directly confronted with it every morning when we clean our teeth.

You can seek out these secrets with your tongue. These are four small points in your mouth — two of them are located on the inside of your cheeks, opposite your upper molars, more or less in the middle. If you explore the area with your tongue you will feel two tiny bumps. If they notice them at all, most people assume they must have bitten themselves in the cheek at some point, but they haven't — these little nubs, which doctors call the parotid

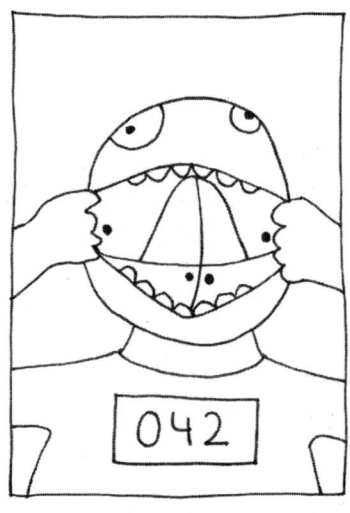

• = papillae

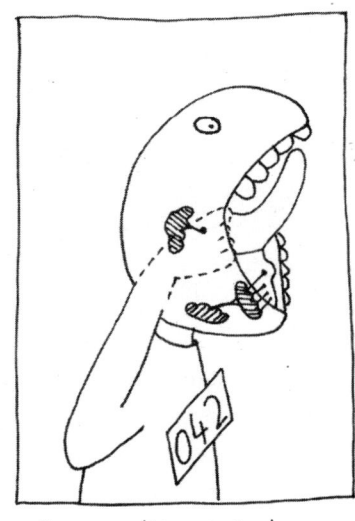

🕮 = salivary glands

papillae, are found in the same position in everybody's mouth. The other two points are lurking beneath your tongue, just to the right and left of the lingual frenulum, the fold of skin connecting the tongue to the floor of your mouth. These four little nubs supply your mouth with saliva.

The papillae in your cheeks secrete saliva whenever it's needed right away — for example, when we eat. The two tiny openings under the tongue secrete saliva continuously. If you could somehow enter these channels, and swim against the tide of saliva, you would eventually reach the main salivary glands. They produce the most saliva — about 0.7 to one litre a day. If you feel upwards from your neck to your cheek, you will notice two soft, round raised areas. May I introduce you? They are the bosses.

The sublingual papillae, those two constant suppliers of saliva, are situated right behind our lower front teeth, which are particularly susceptible to the build-up of tartar. This is because there are substances in our saliva that contain calcium, which are actually only intent on making our teeth harder. But if a tooth is constantly bombarded with calcium, it can be a case of 'too much of a good thing'. Tiny molecules floating innocently by are caught up and 'fossilised' without so much as a by-your-leave. The problem is not the tartar itself, but the fact that it has such a rough surface, affording a much better foothold for bacteria that cause tooth decay and gum disease than does smooth, clean tooth enamel.

But what are fossilising, calcium-containing substances doing in our saliva? Saliva is, basically, filtered blood. The salivary glands sieve the blood, keeping back the red blood cells, which are needed in our arteries, not our mouth. But calcium, hormones, and some products of our immune system enter the saliva from the blood. That explains why each person's saliva is slightly different. In fact, saliva analysis can be used to test for diseases of the immune system, or for certain hormones. The salivary glands can also add extra substances, including those calcium-containing compounds, and even natural painkillers.

Our saliva contains one painkiller that is stronger than morphine. It is called opiorphin, and was only discovered in 2006. Of course, we produce only small amounts of this compound — otherwise we would be spaced out on our own spit all the time. But even a small amount has a noticeable effect, since our mouth is such a sensitive thing. It contains more nerve endings than almost anywhere else in the human body — even the tiniest strawberry seed can drive us crazy if it gets stuck somewhere; we feel every grain of sand in a badly washed salad. A teeny little sore, which we would not even notice if it were on our elbow, hurts like hell and feels monstrously big in our mouth.

Without our salivary painkiller, it would feel even worse! When we chew, we produce more saliva, and with it more of such analgesic substances, which explains why a sore throat often feels better after a meal, and even minor sores in the oral cavity hurt less. It doesn't have to be a meal — even chewing gum provides us with a dose of our oral anodyne. There are even a handful of new studies showing that opiorphin has anti-depressant properties. Is our spit partly responsible for the reassuring effects of comfort eating? Medical research into both pain and depression may deliver the answers in the next few years.

Saliva protects the oral cavity not only from too much pain, but also from too many bad bacteria. That's the job of mucins, for example. Mucins are proteins that form the main constituent of mucous. They help provide hours of fascination and fun for young children who have just found out they can blow bubbles with their own spit. A more useful function is their ability to envelop our teeth and gums in a protective mucin net. We shoot them out of our salivary papillae like Spiderman shoots webs from his wrists. These microscopic nets can catch bacteria before they have a chance to harm us. While the bad bacteria are caught in the net, anti-bacterial substances in our saliva can kill them off.

Like the natural painkillers in our saliva, bactericidal substances are present in our saliva in small concentrations. Our spit is not supposed to disinfect us completely. In fact, we actually need a core team of good little creatures in our mouths. Benign bacteria in the mouth are not totally wiped out by our disinfectant saliva, since they take up space — space that could otherwise be populated by more dangerous germs.

When we are asleep we produce very little saliva. That's good news for those who tend to drool into their pillow — if they produced the full daytime quota of one to one-and-a-half litres during the night, too, the results would not be particularly pleasant. The fact that we produce so little saliva at night explains why many people have bad breath or a sore throat in the morning. Eight hours of scarce salivation mean one thing for the microbes in our mouth: party time! Brazen bacteria are no longer kept in check, and the mucous membranes in our mouth and throat miss their sprinkler system.

That is why brushing your teeth before you go to bed at night and after you get up in the morning is such a clever idea. Brushing at bedtime reduces the number of bacteria in your mouth, leaving fewer partygoers for the all-night bash. Brushing in the morning is like cleaning up after the party the night before. Luckily, our salivary glands wake up at the same time we do in the morning, and start production straight away. Munching on our first piece of toast or performing our morning dental-hygiene duties adds extra stimulation for salivation, and this washes away the nocturnal microbes or transports them down into our stomach, where our gastric juices finally finish them off.

Those who suffer from bad breath during the day may have not managed to remove enough musty-smelling bacteria. Those cunning little critters love to hide out under the newly formed mucin net where the anti-bacterial substances in our saliva cannot get to them. A tongue scraper can help here, but so can chewing gum. It helps stimulate saliva production to swill away those mucin hide-outs. If none of this helps, there is another place where the causes of bad breath can lurk. But more of that later, after we have found out about the second secret place in our mouths.

This place is one of those typical surprises — like when you think you know someone, only to find out they have an unexpected, crazy side to them. The well-coiffed secretary from the city who turns up on the Internet as a fanatical ferret breeder. The heavy-metal guitarist seen buying skeins of yarn, because he finds knitting so relaxing, and it's such a good workout for the fingers. The best surprises come after first impressions have been made, and the same is true of our own tongue. When you look in the mirror and stick out your tongue, you are not seeing it in all its glory. You might well ask how it looks further down, as it is clear to see that it does not just end at the back of your mouth. In fact, the root of the tongue is where things really start to get interesting.

It is home to an alien landscape of pink domes. Those whose gag reflex is not too pronounced can carefully feel the root of their tongue with a finger. When you reach the root, you will notice it gets pretty bumpy back there. The job of these nodules — doctors call them your lingual tonsils — is to investigate everything we swallow. To do this, they pick up tiny particles of anything we eat, drink, or inhale, and draw them into the nodule. Inside, an army of immune cells waits to receive training in how to deal with foreign substances invading from the outside world. They need to learn to leave bits of apple in peace, while attacking anything that might give us a sore throat. So, if you do explore the root of your tongue with your finger, it is not certain who is explorer and who the explored — after all, this area includes some of the most inquisitive tissue in our bodies: immune tissue.

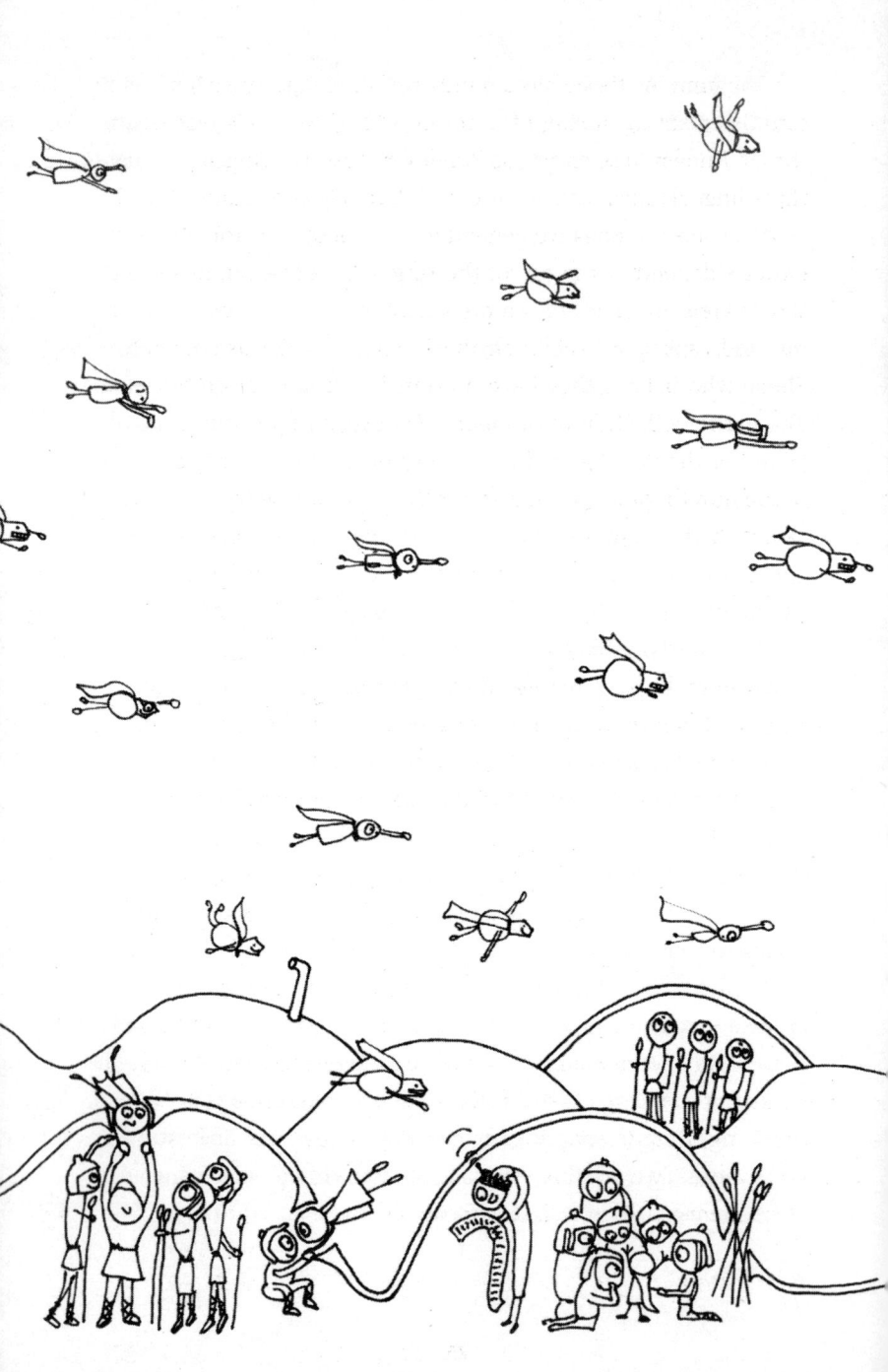

The immune tissue has a number of such inquisitive hotspots; strictly speaking, a ring of immune tissue encircles our entire throat. Known to scientists as Waldeyer's tonsillar ring, it includes those lingual tonsils at the bottom of the circle, the palatine tonsils — these are the ones we generally think of as our 'tonsils' — at either side, and at the top of the ring, where the ear, nose, and throat areas meet, there is more such tissue. (When swollen and infected, especially in children, this is what we know as 'adenoids'.) Those who believe they have no tonsils left are not quite right. The entire collection of tissue in Waldeyer's ring is our 'tonsils'. Whether they are located at the root of the tongue, at the back of the mouth, or at the side of our throat, all these tonsils do the same job: they inquisitively investigate any foreign substance they encounter, and use the information to train the immune system to defend us.

The tonsils — the ones we often have removed — are just not as clever in the way they go about it. Rather than forming bumps, they tend to form deep grooves (to increase surface area), known rather spookily as 'crypts'. Sometimes, too much foreign material can get caught in the crypts, leading to frequent infections. This is a side effect, so to speak, of having over-inquisitive tonsils. So, if the tongue and teeth have been excluded as a cause of a patient's bad breath, the next place to check is the tonsils — if they are still there.

Sometimes, little white stones can be found hiding in the crypts, which smell terrible! Often, people have no idea they are there, and spend weeks trying unsuccessfully to get rid of bad breath or a strange taste in their mouth. No amount of tooth brushing, tongue scraping, or gargling helps. The little stones will eventually work their way out of their hiding places, with no permanent harm done. But you can also take fate into your own

Fig.: *The immune tissue at the base of the tongue, also called the lingual tonsils* hands and, with a little practice, squeeze them out. That done, bad breath problems disappear instantaneously.

The best test to find out whether smelly breath is caused by these little deposits is simply to run a finger or a cotton bud over the tonsils and then sniff it. If it smells unpleasant, it is time to go hunting for tonsil stones. Ear, nose, and throat doctors can also remove them — which is the safer and more convenient option. Those with a strong stomach and a love of barely watchable videos can visit YouTube to see various techniques for squeezing tonsil stones out, and to view some extreme examples. But be warned — they are not for the faint-hearted.

There are also other household remedies for tonsil stones. Some people gargle with salt water several times a day; others swear by fresh, raw sauerkraut from the health-food store; and yet others claim that cutting out dairy products will prevent them from forming at all. There is no scientific basis for any of these remedies. A more thoroughly researched medical question is that of when a tonsillectomy can or should be carried out. The answer turns out to be: not before the age of seven.

That is the age by which we have probably seen it all, or all that is important for our immune cells: being born into a completely unfamiliar world; being kissed and cuddled by Mummy; playing in the garden or the woods; touching animals; having many colds in quick succession; meeting a load of new people at school. And that's about all. By this time, our immune system has finished its schooling, so to speak, and can go to work for us for the rest of our lives.

Before we reach the age of seven, our tonsils are still an important training camp for our immune cells. Building a healthy immune system is not only important for warding off colds; it also has an important part to play in keeping our hearts healthy and in controlling body weight. For example, removing the tonsils of a child younger than seven can lead to an increased risk of obesity.

Why this should be the case is something doctors have not yet found out. However, more and more researchers are now becoming interested in the link between the immune system and body weight. This tonsil-tubbiness-effect can be a boon for underweight children. The associated weight-gain can propel them into the normal weight range. But for all other children, parents are best advised to make sure their offspring eats a healthy, balanced diet after a tonsillectomy.

So the tonsils of children below the age of seven should stay in, unless there is a very good reason for taking them out. If the tonsils are so large that they impede normal breathing or sleeping, for example, the tonsil-tubbiness-effect is secondary. It may seem sweet of our immune tissue to want to defend us so loyally, but in such cases, it does more harm than good. Often, doctors can use lasers to remove only that part of the tonsils which is causing the trouble; they no longer have to leave patients completely tonsil-less. Chronic or repeated infections are a different story altogether. In such cases, our immune cells are kept constantly busy, with no time for a bit of R&R, and that is not good for them if it continues for too long. Whether we are four, seven, or fifty years old, oversensitive immune systems can benefit from saying goodbye to those tonsils.

One example of this is psoriasis sufferers. An over-reaction of the immune system causes itchy skin lesions (often starting at the head), and painful inflammation of the joints. Psoriasis patients also have an above-average vulnerability to sore throats. One possible factor in this is bacteria, which can hide in the tonsils for long periods of time and rankle the immune system from there. For more than thirty years, doctors have described cases of psoriasis patients whose skin condition improved or cleared up entirely following a tonsillectomy. In 2012, this prompted researchers from Iceland and the USA to investigate the phenomenon more closely. They split 29 psoriasis patients who also suffered frequent sore throats into two groups. One group had their tonsils surgically removed; the other didn't. Thirteen of the 15 'detonsilised' patients reported a clear, long-term improvement in their skin. Those still in possession of their tonsils reported little or no change. Some sufferers of rheumatic diseases are also now advised to have their tonsils removed when they are suspected of involvement in the cause of the condition.

Tonsils in or tonsils out — there are good arguments for both. Those forced to bid farewell to their tonsils at an early age need not worry that their immune system has missed an important lesson from the oral cavity. Luckily, there is still all the rest of the tissue at the base of the tongue and back of the throat. And those whose tonsils are still in place need not worry that they have been left with nothing but a trap for bacteria. Many people's tonsillar crypts are rather shallow, and so are less likely to cause problems for their owners. The other parts of Waldeyer's ring are, in fact, very bad at providing a hideout for bacteria. They are constructed differently, and have glands to help them clean themselves regularly.

There is something happening every second in our mouths: salivary papillae shoot out nets of mucin, take care of our teeth, and protect us from the effects of oversensitivity. Our tonsillar ring keeps watch for foreign particles, and uses them to train its immune army. But we would need none of this, if the story didn't continue beyond our mouth. It is simply the gateway to a world where the external becomes internalised.

The Structure of the Gut

Some things turn out to be a disappointment once you get to know them better. Those chocolate wafers from the TV commercial are not lovingly hand-baked by housewives in country dresses; they come from a factory with neon strip-lighting and workers at production lines. School turns out to be much less fun than you thought it would be on the first day. It's 'warts and all' in the backstage area of life, where there is a lot that looks much better from a distance than up close.

That is not the case for the gut, however. Our intestinal tube looks rather odd from a distance. Beyond the mouth, a two-centimetre-wide oesophagus, or gullet, leads down from the throat, misses the top of the stomach, and passes into it somewhere at the side. The right-hand side of the stomach is much shorter than the left, which is why it curls up into a crescent-shaped, lopsided pouch. The small intestine meanders with no particular sense of direction, sometimes to the right, sometimes to the left, for all its seven metres in length until it eventually passes into the large intestine. That's where we find the apparently useless appendix, which seems to be incapable of doing anything except getting infected. The large intestine is also full of bulges. In fact, it looks a little like a sorry attempt to replicate a string of beads. Seen from a distance, the gut is an unsightly, charmless, asymmetrical tube.

So let's forget the view from a distance, and zoom in for a closer look. There is scarcely another organ in our body that becomes increasingly fascinating, the closer you get. And the more you know about the gut, the more beautiful it appears. Let's look at some of those strange structures a little more closely.

The 'gargly' oesophagus

The first thing we notice is that the oesophagus can't aim properly. Rather than taking the shortest route and aiming for the middle of the stomach, it enters the organ on the right-hand side. This is a smart move. Surgeons would call such a connection 'terminolateral'. It may mean taking a little detour, but it's well worth it. When simply walking normally, we tense our abdominal muscles, doubling the pressure in our abdomen with every step we take. When we laugh or cough, for example, that pressure increases by several times. Since the abdomen presses against the stomach from below, it would be a bad idea for the oesophagus to dock directly onto the top end of the stomach. Connected as it is at the side, it has to deal with only a fraction of the pressure. It is thanks to this arrangement that we can take a walk after a heavy meal without having to burp with every step. This clever angle and its closing mechanism are also to thank for the fact that, although a fit of laughter might result in us losing a little control over our outer sphincter and inadvertently letting out a little 'laughing gas', few people have been known to vomit from laughing.

A side effect of this lateral connection is the so-called gastric bubble. This small bubble of air at the top of the stomach can be seen clearly on X-rays. Air rises vertically, after all, and does not search out a side exit. This bubble is the reason that many people find they have to swallow a little air in order to burp. This swallowing motion moves the opening of the oesophagus a little closer to the bubble, and — hey presto! — the burp can make its upward journey to freedom. Those who need to burp while lying down can make the process easier by lying on their left side. So, if

---

Fig.: *In order better to illustrate the stomach bubble, this figure does not show the correct distribution of black and white in a normal X-ray image. Normally, denser material, such as teeth or bone, shows up white, while less dense material, such as the stomach bubble or the air in the lungs, shows up as dark areas.*

Valera Ekimotchev
Birth Date: 1/16/1983
ID: 3782953
Acc No: 7722536

Radiology
Acq. time: 23:13:11

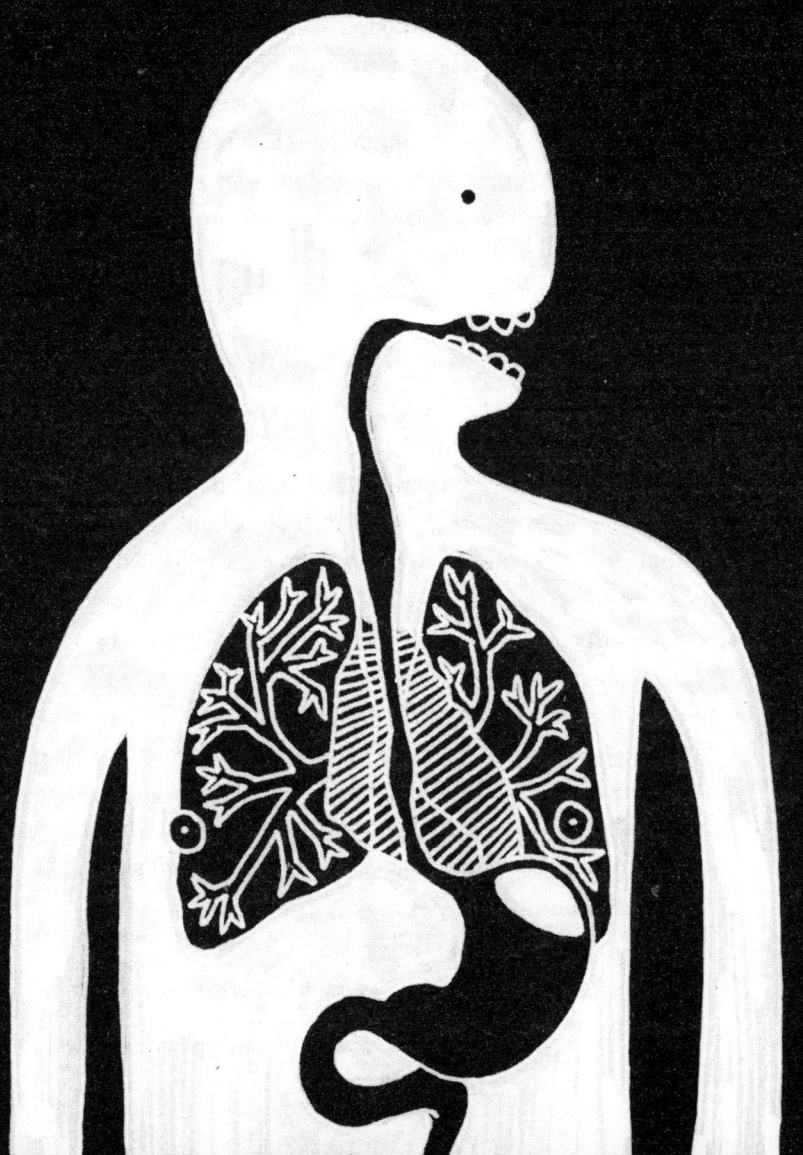

you're kept awake at night by a bloated stomach, and you are lying on your right side, the best thing to do is simply to turn over.

The 'gargly' appearance of the oesophagus is also more beautiful than it seems at first glance. Looking very closely, it can be seen that some muscle fibres run around the oesophagus in a spiral pattern. They are the reason for its 'gargly' motion. If you extend these fibres lengthways, they constrict spirally, like a telephone receiver cable. Bundles of fibres connect the oesophagus to the spinal column. Sitting up straight and looking upwards stretches the oesophagus along its length. This causes it to narrow, in turn allowing it to close more efficiently at each end. That is why sitting or standing up straight can help prevent heartburn after a large meal.

The lopsided stomach pouch

Our stomach sits much higher in our abdomen than we think. It begins just below the left nipple and ends below the bottom of the ribcage on the right. Any pain felt further down than this lopsided little pouch cannot be stomach ache. Often, when people say they have stomach problems, the trouble is actually in the gut. The heart and the lungs sit on top of the stomach. This explains why we find it more difficult to breathe deeply after eating a lot.

An often-overlooked condition is Roemheld syndrome, or bacterial overgrowth, when so much gas collects in the stomach that it presses against the heart and the vagus nerve, which innervates many of our internal organs. Sufferers may display a range of different symptoms, including dizziness and discomfort. In more severe cases, Roemheld syndrome can cause anxiety or difficulty in breathing, and may also lead to severe chest pain that feels like a heart attack. Doctors often write off undiagnosed Roemheld sufferers as overanxious malingerers whose symptoms are all in their minds. A more useful approach would be to ask patients if they have tried burping or passing wind. In the long term, it might be better for such patients to avoid any food that leaves them bloated or flatulent, take measures to restore the balance of the stomach or gut flora, or avoid drinking alcohol to excess. Alcohol can multiply the number of gas-producing bacteria by a factor of up to a thousand. In fact, some bacteria feed on alcohol (which is why rotten fruit tastes alcoholic). With a gut full of busy gas producers, a night on the town can lead to a morning chorus of the pungent kind. So much for the 'alcohol is a disinfectant' argument!

Now let us turn to the stomach's strange shape. One side is much longer than the other, and so the entire organ has to bend double. This creates large folds inside it. The stomach could be called the Quasimodo of the digestive organs. But its misshapen appearance has a deeper meaning. When we take a drink of water, the liquid is able to flow straight down the shorter, right-hand side of the stomach to end up at the entrance to the small intestine. Food, on the other hand, plops against the larger side of the stomach. So, our digestive pouch cunningly separates the substances it still needs to work on, to break them down, from fluids that it can wave straight on through to the next digestive station. Our stomach is not simply *lopsided*; rather, it has two sides with different specialisations. One side copes better with fluids; the other, with solids. Two stomachs for the price of one, so to speak.

The meandering small intestine

The small intestine meanders about in our tummies, twisting and turning for a distance of between three and six metres. If we bounce on a trampoline, it simply bounces along with us. When the plane we're sitting on takes off, it is pressed into the back of the seat like the rest of us. When we dance, it merrily wobbles along to the music, and when stomach ache makes us wince, its muscles 'wince' in a similar way.

There are few people in the world who have seen their own small intestine. Even doctors usually examine only the large intestine when they perform a colonoscopy. But those who do get the rare opportunity of seeing their small intestine, by swallowing a pill-sized camera, are likely to be surprised. Most expect to encounter a gloomy tunnel, but what they see is a very different creature: moist, pink, with a velvety sheen and somehow delicate-looking. Most people do not realise that only the final metre or so of our large intestines has anything to do with faeces — the preceding metres are surprisingly clean (and largely smell-free, incidentally). They faithfully and tastefully take on everything we swallow down to them.

At first sight, the small intestine seems rather more haphazard in its design than our other organs. The heart has its four chambers, the liver has its lobes, veins have valves, and the brain has specialised areas — but the small intestine just wanders aimlessly about in our abdomen. Its true design becomes clear at the microscopic level. We have here a creature that epitomises the phrase 'love of detail'.

Our gut wants to offer us as much surface area as possible. That is why it loves folds. Those include the folds we can see with the naked eye — without them, our small intestine would need to be up to 18 metres long to provide us with enough surface area for our digestion. So, here's to folds! But a perfectionist like the small intestine doesn't stop there. Each square millimetre of the surface contains some thirty tiny finger-like projections, called 'villi' by scientists, which protrude out into the mush of partly digested food — the medical word for which is 'chyme'. The villi's size means they appear as a velvety structure to the unaided human eye. Under the microscope, the little villi look like large waves made out of

Fig.: *Intestinal villi, microvilli and glycocalyxes*

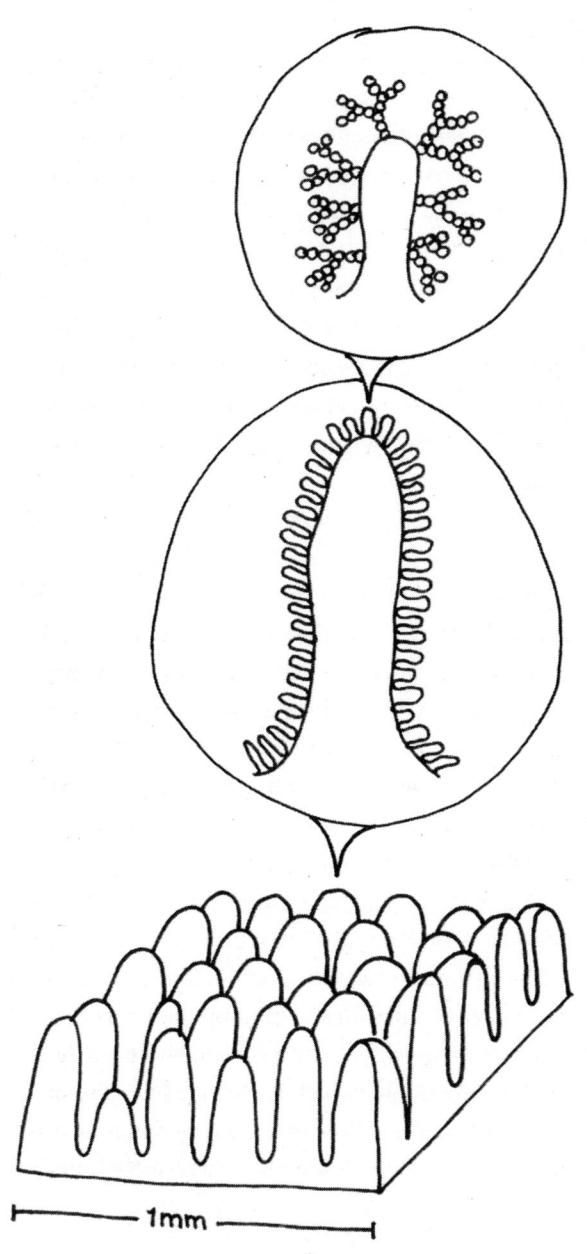

cells. (Velvet looks very similar under a microscope.) Even greater magnification reveals that each and every one of those cells is itself covered with little protrusions — the microvilli — villi on villi, if you like. The microvilli are in turn covered with a velvety meshwork made of countless sugar-based structures that look a little bit like antlers. These are called the glycocalyxes. If all this — the folds, the villi, and the microvilli — were ironed out to a smooth surface, our gut would have to be some seven kilometres in length.

Why does it have to be so huge, anyway? In total, the surface area of our digestive system is about forty times greater than the area of our skin. That seems a little excessive just to deal with a small portion of fries or a single apple. But this is what it's all about inside our bellies: we enlarge ourselves as much as possible in order to reduce anything from outside to the smallest size we can, until it is so tiny that our bodies can absorb it and it eventually becomes a part of us.

We begin that process in the mouth. A bite of an apple sounds like such a juicy idea because, when we take that bite, our teeth burst millions of apple cells like tiny balloons. The fresher the apple is, the more of its cells remain intact — which is why we can tell how fresh the fruit is by its crispness as we bite into it.

Just as we prefer crisp, fresh fruit, we also love hot, protein-rich food. We find steak, scrambled eggs, or fried tofu more appetising than raw meat, slimy eggs or cold bean curd. That's because we have an intuitive understanding of how digestion works. If we swallow a raw egg, it will undergo the same processes in our stomach as it would in the frying pan. The white of the egg turns opaque, the yolk takes on a pastel colour, and both set and become solid. If we were to vomit the raw egg back up after the right amount of time, the results would look like almost perfect scrambled eggs — without any cooking! Proteins react to the heat in the hot pan and the acid in our stomach in the same way — they unfold. That means they no longer possess the clever design features that make them soluble in the liquid of the egg white, so they form solid white lumps. In this state, they can be digested far more easily in the stomach and small intestine. Cooking food saves us the whole first burst of energy required to unfold those proteins, which would otherwise have to be expended by the stomach. By preferring cooked food, the body 'outsources' the first part of the digestive process.

The final breakdown of the food we eat takes place in the small intestine. Right at the start of this part of the gut there is a small opening in its wall. This is the duodenal papilla — similar to the salivary papillae in our mouths, but bigger. It is through this little hole that digestive juices are squirted onto the chyme. As soon as we eat something, the liver and pancreas begin to produce these juices and deliver them to the papilla. They contain the same agents as the washing powder and washing-up liquid you can buy from any supermarket: digestive enzymes and fat solvents. Washing powder is effective in removing stains because it 'digests out' any fatty, protein-rich, or sugary substances from your laundry, with a little help from the movement of the washing-machine drum, leaving them free to be rinsed down the drain with the dirty water. That is more or less the same as what happens in our small intestine. The main difference is that the pieces of protein, fat, or carbohydrates broken down in the intestine, ready to be transported to the bloodstream through the gut wall, are huge by comparison. A bite of an apple is then no longer a bite of apple, but a nutritious pulp made up of billions and billions of energy-rich molecules. Absorbing them all requires a huge surface area — seven kilometres in length are just about enough. That also leaves some space as a safety buffer, in case parts of the gut are temporarily put out of action by infection or gastric flu.

Each individual villus contains a tiny blood vessel — a capillary — that is fed with the absorbed molecules. All the small intestine's blood vessels eventually come together and carry the blood to the liver, where the nutrients are screened for harmful substances and toxins. Any dangerous substances can be destroyed here before the blood passes into the main circulatory system. If we eat too much, this is where the first energy stores are created. The nutrient-rich blood then flows from the liver directly to the heart. There, it receives a powerful push and is pumped to the countless cells of our body. In this way, a sugar molecule can end up in a skin cell in your right nipple, for example, where it is absorbed and then 'burnt' along with oxygen. That releases energy which the cell uses to stay alive, with heat and tiny amounts of water created as by-products. This happens inside so many cells at the same time that the heat produced keeps our body at a constant temperature of 36 to 37 degrees centigrade.

The basic principle underlying our energy metabolism is simple. Nature requires energy to ripen an apple on the tree. We humans then come along and break the apple down into its constituent molecules and metabolise them for energy. We then use the energy released to keep us alive. All the organs that develop out of that embryonic gut tube are able to provide fuel for our cells. Our lungs, for example, do nothing other than absorb molecules with every breath we take. Thus 'breathing in' really means 'taking in nourishment in gaseous form'. A good proportion of our body weight is made from such inhaled atoms, and not from cheeseburgers. Indeed, plants draw the majority of their weight from the air and not from the soil they grow in ... I hope I haven't just inadvertently provided the next dubious 'diet' idea to appear in the women's magazines.

So all our body's organs use up energy, but it is from the small intestine that we start to get some energy back. That explains why eating is such a pleasant pastime. However, we can't expect to feel a burst of energy as soon as we have swallowed the last mouthful of a meal. In fact, many people find they feel tired and sluggish after eating. The food has not yet reached the small intestine —

it is still in the preparatory stages of digestion. We no longer feel hungry because our stomach has been expanded by the food we've eaten. But we feel just as sluggish as we did before the meal, and now we have to come up with the extra energy for all that mixing and breaking down. To achieve this, a large amount of blood is delivered to our digestive organs, and many researchers believe that post-prandial tiredness may also be due to the resulting reduced blood supply to the brain.

One of my professors always dismissed this idea, arguing that if all the blood in our heads were diverted to our stomachs we would be dead, or at least unconscious. Indeed, there are other possible causes of the fatigue that follows eating. Certain messenger chemicals released by the body when we are full can also stimulate the areas of the brain responsible for tiredness. This tiredness is perhaps inconvenient for our brains when we are at work, but the small intestine welcomes it. It works most effectively when we are pleasantly relaxed. It means the optimum amount of energy is available for digestion, and our blood is not full of stress hormones. The phlegmatic after-lunch reader is a more efficient digester than the stressed-out office executive.

The unnecessary appendix and the bulgy large intestine

There are nicer things in life than lying on an examining table at the doctor's, with one thermometer in your mouth and another in your behind. But that used to be the standard examination in cases of suspected appendicitis. A significantly higher temperature down below than in the mouth was a major indication. Modern doctors no longer need to rely on temperature differences to diagnose appendicitis. Important symptoms are fever, in combination with pain below and to the right of the belly button (the position of the appendix in most people).

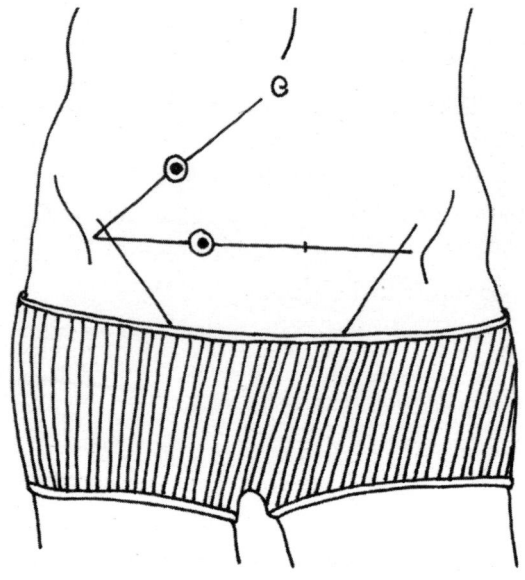

Often, pressing that side of the lower abdomen will cause pain; curiously, pressing the other side will relieve it. As soon as pressure on the left-hand side is released — ouch! This is because our abdominal organs are surrounded by a supporting fluid. When pressure is applied to the left-hand side, extra support fluid is pushed over to the right, where it provides additional cushioning for the inflamed gut, which relieves pain. Other signs of appendicitis are pain when raising the right leg against a resistant pressure (get someone to push against it), lack of appetite, or nausea.

Our appendix, officially known as the vermiform, or 'worm-shaped' appendix, has a reputation for being useless. Looking like a deflated balloon of the kind that children's party entertainers twist into animal shapes, the appendix is not only too small to deal with chyme, it is also positioned in a location that partly digested food hardly ever reaches. It is just below the junction between the small and large intestines, and is completely bypassed. This is a creature that can only look on from below as the world continues on its way above. Those of you who remember the bumpy landscape in our mouths might have an idea about what its true function might be. Although far removed from the rest of its kind, the appendix is part of the tonsillar immune tissue.

Our large intestine takes care of things that cannot be absorbed in the small intestine. For that reason, it does not have the same velvety texture. It would simply be a waste of energy and resources to fill this part of the gut with absorbent villi. Instead, this is the home of most of our gut bacteria, which can break down the last nutritious substances for us. And our immune system is very interested in these bacteria.

The vermiform appendix couldn't be better placed! It is far enough away so as not to be bothered by all the digestive business going on above it, but close enough to monitor all foreign microbes. Although the walls of the large intestine include large deposits of immune cells, the appendix is made almost entirely of immune tissue. So, if a bad germ comes by, it is surrounded. However, this also means that everything around it can become infected — 360° panoramic inflammation, so to speak. If this inflammation causes the appendix to swell, the little tube has problems sweeping itself clean of those bad germs — leading to one of the more than 45,000 appendectomies carried out every year in the United Kingdom alone (and over 25,000 in Australia).

However, that is not the only function of the appendix. It leaves only good germs alive and attacks anything it sees as dangerous, and this also means that a healthy appendix acts as a storehouse of all the best, most helpful bacteria. This was discovered by American researchers Randall Bollinger and William Parker in 2007. Its practicality comes into play after a heavy bout of diarrhoea. That will often flush away many of the typical gut microbes, leaving the terrain free for other bacteria to settle. That should not be left to chance. And this is when, according to Bollinger & Parker, the appendix team steps in and spreads out protectively through the entire large intestine.

Germany, where I live, is not a region that contains many pathogens which cause diarrhoea. We may pick up a gastro-intestinal flu bug every now and then, but our environment teems with far fewer dangerous microbes than in India or Spain, for example. So you could say that we do not need our appendix as urgently as the people in those regions do. That means no one at home who has undergone an appendectomy, or is about to face one, should be all too worried. The immune cells in the rest of the large intestine may not be quite so closely packed; but, in total, they are many times more numerous than those in the appendix, and are competent enough to take on the job. Anyone who wants to take no chances after a bout of diarrhoea can buy good bacteria at the pharmacy to repopulate his or her gut.

So now, I hope, it is clear why we have an appendix. But what's the purpose of the large intestine that it is appended to? Nutrients have already been absorbed, there are no villi here, and what does our gut flora even want with indigestible leftovers, anyway? Our large intestine does not wind about like its smaller counterpart. It surrounds our small intestine on the outside, like a plump picture frame. And it would not take exception to being called 'plump' — it simply needs more room to do its job.

'Waste not, want not' may sound hackneyed today, but for past generations it was a way to survive lean times. And it is also the motto of our large intestine. It takes its time with all the leftovers, and digests them thoroughly. The small intestine can get on with processing the next meal, or even the next two, in the meantime, without affecting the large intestine's work. It doggedly processes leftovers for sixteen hours or so. In doing so, it makes available substances that would have been lost if the gut were more hurried. They include important minerals like calcium, which can only be absorbed properly here. The careful cooperation of the large intestine and its flora also provides us with an extra helping of energy-rich fatty acids, vitamin K, vitamin B12, thiamine (vitamin B1), and riboflavin (vitamin B2). Those substances are useful for many things — for example, to help our blood clot properly, to strengthen our nerves, or to prevent migraines. In the final metre of the large intestine, our water and salt levels are finely tuned: not that I'm recommending a taste test, but the saltiness of our faeces always remains the same. This fine balancing act saves the body an entire litre of fluids, which we would have to make up by drinking an additional litre per day.

As with the small intestine, all the treasures absorbed by the large intestine are transported first to the liver for checking before entering the main blood system. The final few centimetres of the large intestine, however, do not send their blood to the detoxifying liver; blood from their vessels goes straight into the main circulatory system. This is because, generally, nothing more is absorbed in this section, simply because everything useful has already been removed. But there is one important exception: any substances contained in a medical suppository. Suppositories are able to contain much less medication than pills, and still take effect more quickly. Tablets and fluid medication often have to contain large doses of the active agent because much of it is removed by the liver before it even reaches the area of the body it is meant to act on. That is, of course, less than ideal, since the substances recognised by the liver as 'toxins' are the reason we take the medicine in the first place. So, if you want to do your liver a favour and still need to take fever-reducing or other medication, make use of the short cut via the rectum and use a suppository. This is an especially good idea for very young or very old patients.

What We Really Eat

The most important phase of our digestion takes place in the small intestine, where the maximum surface area meets the maximum reduction of our food down to the tiniest pieces. This is where the key decisions are made. Can we tolerate lactose? Is this food good for our health? Which food causes allergic reactions? Here, in this final stage of breakdown, our digestive enzymes work like tiny pairs of scissors. They snip away at our food until it shares a lowest common denominator with our cells. Canny as ever, Mother Nature here makes use of the fact that all living things are made out of the same basic ingredients: sugar molecules, amino acids, and fats. Everything we eat comes from living things — at this biological level, there is no difference between an apple tree and a cow.

Sugar molecules can be linked to form complex chains. When that happens, they no longer taste sweet, and we know them as the carbohydrates we find in bread, pasta, or rice. After that piece of toast you ate for breakfast has undergone the snipping of the enzyme scissors, the final product is the same number of sugar molecules as a couple of spoonfuls of refined household sugar. The only difference is that household sugar does not require so much work from our enzymes, as it is already broken down into such small pieces when it arrives in the small intestines that it can be absorbed directly into the bloodstream. Eating too much pure sugar at once really does make our blood sweeter for a while.

The sugar contained in white toast is digested relatively quickly by our enzymes. With wholegrain bread, everything moves at a much more leisurely pace. Such bread contains particularly complex sugar chains, which have to be broken down bit by bit. So, brown bread is not a sugar explosion, but a beneficial sugar store. Incidentally, our bodies have to work much harder to restore a healthy balance if a sugar onrush comes suddenly. It pumps out large amounts of various hormones — most importantly, insulin. The result is that we rapidly feel tired again, once this special operation is over. If it doesn't enter the system too quickly, sugar is an important raw material for our bodies. It is used as fuel for our cells, like heat-giving firewood, or to build sugar structures for use in our bodies, such as the antler-like glycocalyxes attached to our gut cells.

Despite the problems, our bodies love sugary, sweet treats; they save the body work, since sugar can be taken up more quickly. The same is true of warm proteins. In addition, sugar can be turned into energy extremely quickly, and our brains reward us for a rush of rapid energy by making us feel good. However, there is one problem: never before in the history of humankind have we been faced with such a huge abundance of readily available sugar. Some 80 per cent of the processed foods found on the shelves of modern-day American supermarkets contain added sugar. On an evolutionary scale, then, we could say that our species has just discovered the secret stash of sweets at the back of the cupboard, and it keeps returning to binge on the booty before collapsing on the couch with tummy ache and a sugar shock.

Even though we know intellectually that too much snacking is bad for us, we can't really blame our instincts for encouraging us to grab every opportunity for a treat. When we eat too much sugar, our bodies simply store it away for leaner times. Quite practical, really. One way the body does this is by relinking the molecules to form long, complex chains of a substance called glycogen, which is then stored in the liver. Another strategy is to convert the excess sugar into fat and store it in fatty tissue. Sugar is the only substance our body can turn into fat with little effort.

Glycogen reserves are soon used up — just about the time during your run when you notice the exercise is suddenly much harder work. That is why nutritional physiologists say we should do at least an hour's exercise if we want to burn fat. It is not until we pass through that first energy dip that we start to tap into those fine reserves. We might find it annoying that our paunch isn't the first to go, but our body is deaf to such complaints. The simple reason for this is that human cells adore fat.

Fat is the most valuable and efficient of all food particles. The atoms are so cleverly combined that they can concentrate twice as much energy per gram as carbohydrates or protein. We use fat to coat our nerves — just like the plastic on an electric cable. It is this coating that makes us such fast thinkers. Some of the most important hormones in our body are made out of fat, and every single one of our cells is wrapped in a membrane made largely of fat. Such a special substance must be protected, and not squandered at the first sign of physical exertion. When the next period of famine comes — and there have been many over the aeons — every gram of fat in that paunch is a life-insurance policy.

Our small intestine also knows the special value of fat. Unlike other nutrients, it cannot be absorbed straight into the blood from the gut. Fat is not soluble in water — it would immediately clog the tiny blood capillaries in the villi of the gut, and float on top of the blood in larger vessels, like the oil on spaghetti water. So fat must be absorbed via a different route: the lymphatic system. Lymphatic vessels are to blood vessels as Robin is to Batman. Every blood vessel inside the body is accompanied by a lymphatic vessel — even each tiny capillary in the small intestine. While our blood vessels are thick and red, and heroically pump nutrients to our tissues, the lymphatic vessels are thin and transparent-whitish in colour. They drain away fluid that is pumped out of our tissue, and transport the immune cells which ensure that everything is as it should be throughout the body.

Lymphatic vessels are so slight because they do not have muscular walls like our blood vessels. Often, they work just by using gravity. That explains why we sometimes wake up in the morning with swollen eyes. Gravity is not very much help when you are lying down. The tiny lymphatic vessels in our face are nicely open, but it is only when we get up and gravity kicks in that the fluid transported there during the night by our blood vessels can flow back down. (The reason our lower legs do not fill up with fluid after a long day on our feet is that our leg muscles squeeze the lymphatic vessels every time we take a step, and that squeezes the fluid — known as 'lymph' in medical circles — upwards.) The lymphatic system appears to be an under-appreciated weakling everywhere in the body, except in the small intestine. This is its time to shine! All the body's lymph vessels converge in an impressively thick duct where all the digested fat can gather without the risk of clogging.

It's well known that doctors like to show off their Latin skills, and so they give this vessel the mighty-sounding name *Ductus thoracicus*. It sounds almost as if it means to say 'Hail Ductus! Teach us why noble fat is so important and evil fat so bad!' Shortly after we eat a fatty meal, there are so many tiny fat droplets in our *Ductus* that the lymph fluid is no longer transparent, but milky white. When the fat has gathered in the *Ductus* — or thoracic duct — it skirts the belly, passes through the diaphragm and heads via a small section of vein, straight for the heart. (All the fluid drained from our legs, eyelids, and gut ends up here.) So, whether it's extra virgin olive oil or cheap chip fat, it all goes straight into the heart. There is no detoxing detour via the liver — as there is for everything else we digest.

Detoxification of dangerous, bad fat takes place only after the heart has given it a powerful push to pump it through the system, and the droplets of fat happen to end up in one of the blood vessels of the liver. The liver contains quite a large amount of blood, and so the probability is high that such a meeting will take place sooner

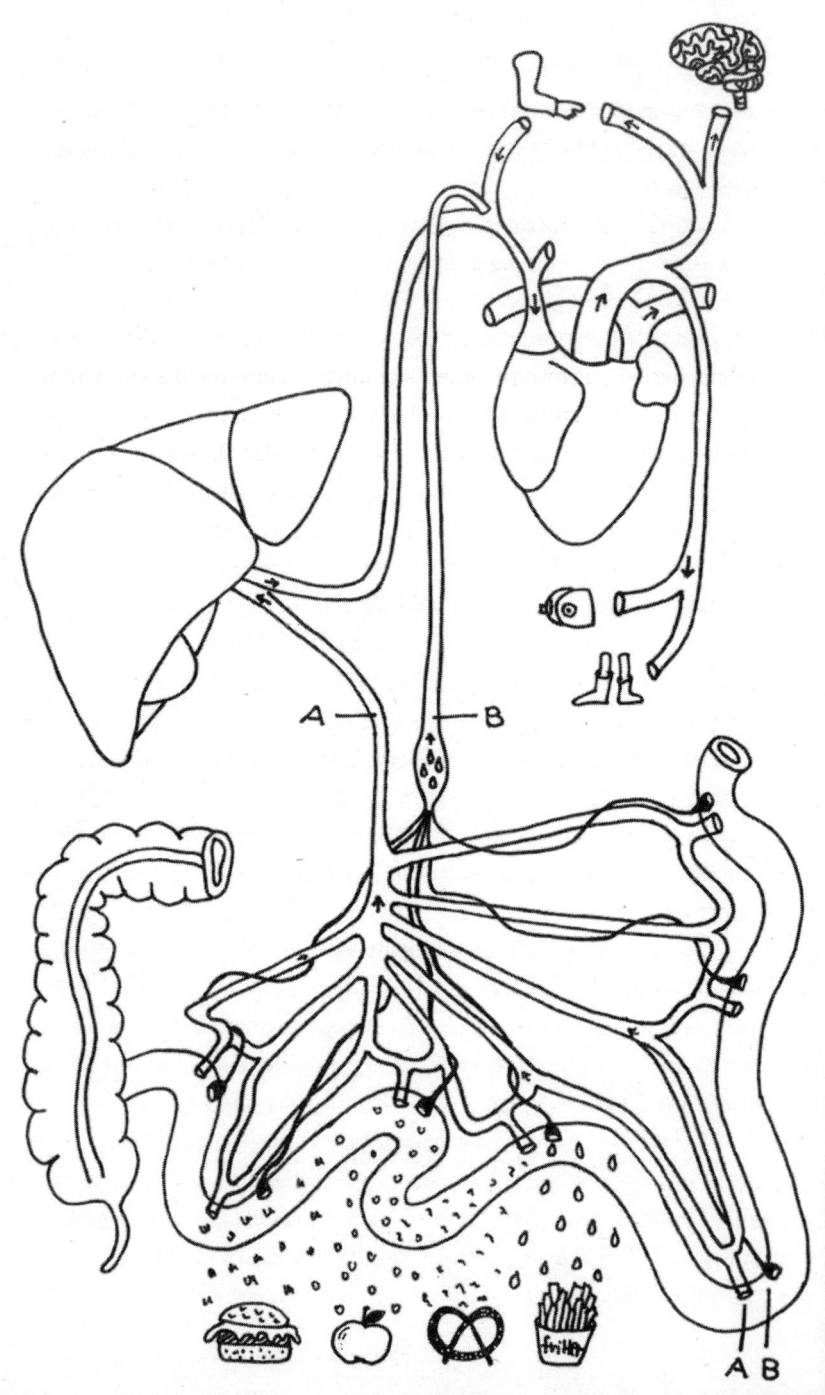

rather than later — but before that happens, our heart and our blood vessels are at the mercy of everything that McDonald's and similar chains have been able to get hold of at the lowest purchase price.

Just as bad fat can have a negative effect, good fat can work wonders. Those who are prepared to spend a little extra on cold-pressed (extra virgin) olive oil will be dunking their baguette in a soothing balm for their heart and blood vessels. Many studies have been carried out into the effects of olive oil, and results show that it can protect against arteriosclerosis, cellular stress, Alzheimer's, and eye disease (such as macular degeneration). It also appears to have beneficial effects on inflammatory diseases such as rheumatic arthritis, and to help in protecting against certain kinds of cancer. Of particular interest to those fighting excessive weight is that olive oil also has the potential to help get rid of that spare tyre. It blocks an enzyme in fatty tissue — known as fatty acid synthase — that likes to create fat out of spare carbohydrates. And we are not the only ones who benefit from the properties of olive oil: the good bacteria in our gut also appreciate a little pampering.

Good-quality olive oil costs a little bit more. However, it tastes neither greasy nor rancid, but rather green and fruity, and sometimes leaves a peppery feeling in your throat after you swallow it. This is due to the tannins it contains. If this description sounds too abstract, simply try out various oils to find the best, using the different quality seals as a guide.

But merrily drizzling your olive oil into the pan for frying is not such a good idea, as heat can cause a lot of damage. Hotplates are great for frying up steaks or eggs, but not for oily fatty acids, which can be chemically altered by heat. Cooking oil, or solid fats such as butter or hydrogenated coconut oil, should be used for frying.

Fig.: *A = Blood vessels pass through the liver and then on to the heart.*
B = Lymphatic vessels go straight to the heart.

They may be full of the much frowned-upon saturated fats, but are much more stable when exposed to heat.

Fine oils are not only sensitive to heat; they also tend to capture free radicals from the air. Free radicals do a lot of damage to our bodies, because they don't actually like being free, much preferring to bond with other substances. They can latch on to almost anything — blood vessels, facial skin, or nerve cells — causing inflammation of the blood vessels (vasculitis), ageing of the skin, or nerve disease. That's why you should always close the bottle or container of olive oil carefully after using it, and keep it in the fridge.

The animal fats found in meat, milk, and eggs, for example, contain far more *arachidonic acid* than vegetable fats. *Arachidonic acid* is converted in our bodies into neurotransmitters involved in the sensation of pain. Oils such as rapeseed (canola), linseed, or hempseed oil, on the other hand, contain more of the anti-inflammatory substance alpha-linolenic acid, while olive oil contains a substance with a similar effect, called *oleocanthal*. These fats work in a similar way to ibuprofen or aspirin, but in much smaller doses. So, although they are no help if you have an acute headache, using these oils regularly can help those who suffer from inflammatory disease, regular headaches, or menstrual pain. Sometimes, pain levels can be reduced somewhat simply by taking care to eat more vegetable fat than animal fat.

However, olive oil should not be seen as a panacea for all skin and hair conditions. Dermatological studies have shown that pure olive oil can even irritate the skin slightly, and that using olive oil as a hair treatment leaves it so oily that the amount of washing required to remove it negates any possible beneficial effects.

It is easy to overdo it with fats inside our bodies, too. Large amounts — of either good or bad fat — are simply too much for the body to deal with. It's comparable to smearing too much moisturiser on your face. Nutritional physiologists recommend we get between 25 and a maximum of 30 per cent of our daily energy requirement from fat. This works out at an average of 55 to 66 grams of fat a day — larger, more athletic types may consume a little more; smaller, more sedentary types should try to consume a little less. This means if you eat just one Big Mac, you will have conveniently covered half your daily requirement of fat. The only question is: With what kind of fat? A chicken teriyaki sandwich from the fast-food chain Subway contains only 2 grams ... how you consume the other 53 grams is then entirely up to you.

Having examined carbohydrates and fats, there is just one more nutritional building block to consider. It is probably the least familiar: amino acids. It seem strange to imagine, but both tofu, with its neutral-to-nutty taste, and salty, savoury meat are made up of lots of tiny acids. As with carbohydrates, these tiny building blocks are linked in chains. This is what gives them their different taste, and a different name — proteins. Digestive enzymes break down these chains in the small intestine and then the gut wall nabs the valuable components. There are twenty of these amino acids, and an infinite variety of ways they can be linked to form proteins. We humans use them, for example, to build our DNA, the genetic material contained in every new cell we produce every day. The same is true of other living things, both plants and animals. That explains why everything nature produces that we can eat contains protein.

However, maintaining a healthy meat-free diet that does not lead to nutritional deficiencies is a more elaborate business than most people think. Plants construct different proteins than animals do, and they often use so little of a given amino acid that the proteins they produce are known as 'incomplete'. When our body tries to use these to make the amino acids it needs, it can continue to build the chain only until one of the amino acids runs out. Half-finished proteins are then simply broken down again, and we excrete the tiny acids in our urine, or recycle them in our bodies. Beans lack the amino acid *methionine*; rice and wheat (and its derivative meat substitute, seitan) lack *lysine*; and maize, in fact, is deficient in two amino acids: *lysine* and *tryptophan*. But this does not spell the final triumph of the meat-eaters over meat-avoiders: vegetarians and vegans simply have to make sure they eat a varied diet.

Beans may be lacking in *methionine*, but they are packed with *lysine* — so that a wheat tortilla with refried beans and a yummy filling will provide all the amino acids the body needs for healthy protein production. Vegetarians who eat cheese and eggs can compensate for incomplete proteins that way. For centuries in many countries around the world, people have intuitively eaten meals made up of foodstuffs that complement each other: rice and beans, pasta with cheese, pita bread and hummus, or peanut butter on toast. Nutritionists used to believe that each meal must include a combination of different foodstuffs. Today we know that that is not necessary. As long as we 'mix it up' a little here and there, our bodies will have no problem with a meat-free diet. There are plants that do contain all the necessary amino acids in the necessary quantities: soya and quinoa are two, but others include amaranth, spirulina, buckwheat, and chia seeds. Tofu has a well-deserved reputation as an alternative to meat — with the caveat that increasing numbers of people are developing allergic reactions to it.

Allergies and Intolerances

One theory about the origin of allergies begins with the digestive processes in the small intestine. If we fail to break down a protein into its constituent amino acids, tiny bits of it will remain. Under normal circumstances, they simply don't make it into our bloodstream, and that causes no problem. However, hidden power often lies in the most inconspicuous places — in this case, in the lymphatic system. Those tiny particles can enter the lymphatic system, embedded in fat droplets, and once there, they attract the attention of ever-vigilant immune cells. When they discover a tiny particle of peanut in the lymphatic fluid, for example, they naturally attack it as a foreign body.

The next time they encounter a peanut particle, they are better prepared to deal with it and can attack it more aggressively. And so it goes on, until we reach the stage where just putting a peanut in our mouth causes our immune cells to whip out the big guns straight away. The result is increasingly severe allergic reactions, such as extreme swelling of the face and tongue. This explanation applies to allergies caused by foods that are both fatty and rich in protein, such as milk, eggs, and, most commonly, peanuts. There is a simple reason why almost no one is allergic to greasy bacon, for example. We are made of meat ourselves, and so we generally have few problems digesting it.

Coeliac disease and gluten sensitivity

Allergies that develop in the small intestine are not limited to fats. Allergens such as prawns, pollen, or gluten, for example, are not fat-bombs in themselves, and people who eat a fatty diet do not necessarily suffer from more allergies than others. Another theory about how allergies develop is this: the wall of our gut can become temporarily more porous, allowing food remains to enter the tissue of the gut and the bloodstream. This is the theory under most scrutiny from researchers who are interested in gluten — a protein found in wheat and related grains.

It is not the case that grains like to be eaten by us. What plants really want is to reproduce — and then along we come, and eat their children. Instead of creating an emotional scene, plants respond by making their seeds slightly poisonous. That sounds much more drastic than it is — neither side is going to lose much sleep over a few guzzled wheat grains. The arrangement means humans and plants both survive well enough. But, the more danger a plant senses, the more poisonous it will make its seeds. Wheat, in particular, is such a worrier because it has only a very short window of opportunity for its seeds to grow and carry on the family line. With such a tight schedule, nothing must be allowed to go wrong. In insects, gluten has the effect of inhibiting an important digestive enzyme. A greedy grasshopper might be put off by a little stomach ache after eating too much wheat, and that is to the benefit of both plant and animal.

In humans, gluten can pass into the cells of the gut in a partially undigested state. There it can slacken the connections between individual cells. This allows wheat proteins to enter areas they have no business being in. That, in turn, raises the alarm in our immune system. One person in a hundred has a genetic intolerance to gluten (coeliac disease), but a considerably higher proportion suffer from gluten sensitivity.

In patients with coeliac disease, eating wheat can cause serious infections or damage to the villi of the gut wall, for example, but it can also damage the nervous system. Coeliac disease can cause diarrhoea, and failure to thrive in children, who may show reduced growth or winter pallor. The tricky thing about coeliac disease is that it can appear in more or less pronounced forms. Those with more subtle forms may live with the symptoms for years without realising it. They may have the occasional belly ache, or their doctor might discover signs of anaemia during routine blood tests. Currently, the most effective treatment is a lifelong gluten-free diet.

Gluten sensitivity, by contrast, is not a sentence to a life of gluten avoidance. Those with this condition can eat wheat without risking serious damage to their small intestine, but they should enjoy wheat products in moderation — a little bit like our friend the greedy grasshopper. Many people notice their sensitivity when they swear off gluten for a week or two and see an improvement in their general wellbeing. Suddenly, their digestive problems or flatulence clear up, or they have fewer headaches or less painful joints. Some people find their powers of concentration improve, or that they are less plagued by tiredness or fatigue. Researchers began exploring gluten sensitivity in any detail only recently. Currently, the diagnostic picture can be summarised as follows: symptoms improve when a gluten-free diet is introduced, although tests for coeliac disease show negative. The villi are not inflamed or damaged, but eating too much bread still appears to have an unpleasant effect on the immune system.

The gut can also become porous for a short time after a course of antibiotics, after a heavy bout of drinking (alcohol), or as a result of stress. Sensitivity to gluten resulting from these temporary causes can sometimes look the same as the symptoms of a true gluten intolerance. In such cases, it can be helpful to avoid gluten for a time. A thorough medical examination and the detection of certain molecules on the surface of the blood corpuscles are important for a definitive diagnosis. Alongside the familiar blood groups A, B, AB, and O, there are many other indicators for categorising human blood, including what doctors call DQ markers. Those who do not belong to group DQ2 or DQ8 are extremely unlikely to have coeliac disease.

Lactose intolerance and fructose intolerance

Lactose intolerance is not an allergy or a real intolerance at all, but a deficiency. But it, too, results from a failure to break down certain nutrients completely into their component parts. Lactose is found in milk, and is derived from two sugar molecules that are linked by chemical bonds. The body requires a digestive enzyme to break that bond; but, unlike other enzymes, this one does not come from the papilla. The cells of the small intestine secrete it themselves on the tips of their tiny little villi. Lactose breaks down when it comes into contact with the gut wall (and the enzyme), and the resulting single sugars can then be absorbed. If the enzyme is missing, similar problems arise to those caused by gluten intolerance or gluten sensitivity, including belly ache, diarrhoea, and flatulence. Unlike in coeliac disease, however, no undigested lactose particles pass through the gut wall. They simply move on down the line, into the large intestine, where they become food for the gas-producing bacteria there. The resulting flatulence and other unpleasant symptoms are, so to speak, votes of thanks from extremely satisfied, overfed microbes. Although the results can be unpleasant, lactose intolerance is far less harmful to health than undiagnosed coeliac disease.

Every human being has the genes needed to digest lactose. In extremely rare cases, problems with lactose digestion can occur from birth. Such new-borns are unable to digest their mother's milk, and drinking it causes severe diarrhoea. In 75 per cent of the world's population, the gene for digesting lactose slowly begins to switch off as they get older. This is not surprising, as by then we are no longer reliant on our mothers' milk, or formula milk, to nourish us. Outside of Western Europe, Australia, and the United

States, adults who are tolerant to dairy products are a rarity. Even in my part of the world, supermarket shelves are increasingly full of lactose-free products. Recent estimates say one person in every five in Germany is lactose intolerant. The older a person, the greater the probability that she will be unable to break down lactose — although very few sixty-year-olds would dream of blaming their daily glass of milk or that delicious dollop of cream for a bloated stomach or a little bit of diarrhoea.

However, lactose intolerance does not mean you must cut out milk products altogether. Most people have enough lactose-splitting enzymes in their gut, but their activity is somewhat reduced — down to about 10 to 15 per cent of their initial level, let's say. So if you notice your tummy feels better when you don't drink that glass of milk, you can simply use trial and error to find out just how much your body can deal with, and how much dairy produce it takes to make the problems come back. A bit of cheese or a splash of milk in your tea or coffee will usually be fine, as will the occasional milk pudding or cream filling in your cake.

The most common food intolerance in Germany is a similar one. A third of all Germans have trouble digesting the fruit sugar fructose. In fact, a common children's counting-out rhyme in Germany, comparable to the 'eeny, meeny, miny, moe' of the English-speaking world, translates as: 'ate cherries, drank water, got tummy ache, went to hospital'! Fructose intolerance can be the result of a severe, congenital inability to metabolise fruit sugar, which causes patients' digestive systems to react even to the tiniest amounts of the substance. But most people affected by fructose intolerance actually have a condition more accurately described as fructose malabsorption, and only experience problems when exposed to large amounts of the sugar. When fructose is described on food packages as 'fruit sugar', consumers often assume it is a healthier, more 'natural' option. This explains why food manufacturers choose to sweeten their products with pure fructose, and consequently why our digestive systems are exposed to more of this type of sugar than ever before.

An apple a day would not present a problem to most people who are fructose intolerant, if it weren't for the fact that the ketchup on their fries, the sweetening agent in their breakfast yoghurt, and the can of soup they heated up for lunch all also contain added fructose. Some types of tomato are specially bred to contain large amounts of this sugar. Furthermore, globalisation and air transport mean that many Western consumers are now exposed to a previously unheard-of overabundance of fruit. Pineapples from the tropics nestle on our supermarket shelves in the middle of winter, next to fresh strawberries from the greenhouses of Holland, and some dried figs from Morocco. So, what we label a food intolerance may in fact be nothing more than the reaction of a healthy body as it tries to adapt within a single generation to a food situation that was completely unknown during the millions of years of our evolution.

The mechanism behind fructose intolerance is different once again from that involved in the digestion of gluten or lactose. The cells of people with congenital fructose intolerance contain fewer fructose-processing enzymes. That means fructose may gather in their cells, where it can interfere with other processes. Fructose intolerance that appears later in life is thought to be caused by a reduced ability of the gut to absorb fruit sugars. Such patients often have fewer transporters (called GLUT-5 transporters) in their gut wall. When they ingest even a small amount of fruit sugar — for example, by eating a pear — their limited transporters are overwhelmed and, as with lactose intolerance, the sugar from the pear ends up feeding the flora of the large intestine. However, some researchers question whether a lack of sufficient transporters really is the cause of this problem, since even those without the condition pass some of the fructose they eat into the large intestine undigested (especially after eating very large amounts of it). The problems experienced by such people may be due to an imbalance in their gut flora. When they eat a pear, the extra sugar is gobbled up by the gut bacteria squad, which then produces rather unpleasant symptoms. Of course, the more ketchup, canned soup, or sweetened yoghurt they have eaten, the worse their troubles will be.

Such a fructose intolerance can also affect our mood. Sugar helps the body absorb many other nutrients into the bloodstream. The amino acid *Tryptophan* likes to latch on to fructose during digestion, for example. When there is so much fructose in our gut that most of it cannot be absorbed into the blood, and we lose that sugar, we also lose the *Tryptophan* attached to it. *Tryptophan*, for its part, is needed by the body to produce *serotonin* — a neurotransmitter which gained fame as the 'happiness hormone' after it was discovered that a lack of it can cause depression. Thus, a long-unrecognised fructose intolerance can lead to depressive disorders. General practitioners and family doctors are only now beginning to include this knowledge in their diagnostic toolkit.

This begs the question of whether a diet that includes too much fructose can also affect our mood, even in the absence of an intolerance. For more than 50 per cent of people, eating 50 grams of fructose or more per day (equivalent to five pears, eight bananas, or about six apples) will overtax their natural transporters. Eating more than that can lead to health problems such as diarrhoea, tummy aches, flatulence, and, over longer periods, depressive disorders. The fructose intake of the average American is currently 80 grams a day. Our parents' generation, consuming just honey on their toast, far fewer processed foods, and a normal amount of fruit, took in only around 16 to 24 grams a day.

Serotonin not only puts us in a good mood, it is also responsible for making us feel pleasantly full after a meal. Snack attacks or constant grazing on snacks may be a side effect of fructose intolerance, if they are accompanied by other symptoms, such as tummy aches. This is also an interesting hint for all diet-conscious salad eaters, since many salad dressings found on our supermarket shelves or at fast-food outlets now contain fructose-glucose syrup (often known as corn syrup in the United States). Studies have shown that this syrup can suppress the hormone that makes us feel full (*leptin*), even in people who are not fructose intolerant. A salad containing the same amount of calories but with home-made vinaigrette or yoghurt dressing will keep you feeling full for longer.

Like everything else in the world, food production is constantly changing. Sometimes those changes are good for us; sometimes they are bad. Curing was once the state-of-the-art way of ensuring that people would not be poisoned by rotten meat. For centuries, it was common practice to cure meats and sausages with large quantities of nitrite salts. This gives them a pinkish-red 'fresh' colour, and explains why products such as ham, salami, tinned pork, or gammon do not turn the same brown-grey colour in the frying pan as an unprocessed chop or steak. The use of nitrites for food preservation has been highly regulated since the 1980s, due to concerns about their possible negative effects on human health. In Europe, sausage and cold-meat products must now contain no more than 100 milligrams (a milligram is one one-thousandth of a gram) of nitrite salt per kilogram of meat — and rates of stomach cancer have fallen considerably since these regulations were introduced. This shows that what had once been a very sensible meat-preserving technique was in drastic need of correction. Today, canny butchers mix large amounts of vitamin C with small amounts of nitrite, to cure their meats safely.

A similar modern reassessment of ancient practices may be in order in the case of wheat, milk, and fructose. It is good to include these foodstuffs in our diet, since they contain valuable nutrients — but it may be time to reassess the quantities we consume. While our hunter-gatherer ancestors ate up to five hundred different local roots, herbs, and other plants in a year, a typical modern diet includes seventeen different agricultural plant crops, at most. It is not surprising that our gut has a few problems with a dietary change of that scale.

Digestive issues divide society into two groups: those who worry about their health and pay great attention to their nutrition; and those who get annoyed by the fact that they can no longer invite a group of friends around for a meal without having to go shopping at the pharmacy. Both groups are right. Many people err on the side of caution after hearing from their doctor about a food intolerance, and then noticing that they do feel better if they avoid certain foods. They might decide to cut out fruit, wheat, or dairy products, and then often act as if they were poisonous. In fact, most people react to excessive amounts of these foodstuffs without being genetically intolerant to them. Most have enough enzymes to process a small portion of creamy sauce, the occasional pretzel, or a fruit pudding.

However, this does not mean that real intolerances should be ignored. We do not need to swallow every new development in our food culture blindly. Wheat products for breakfast, lunch, and dinner; fructose in practically all processed foods; or milk products long after weaning — it is not surprising that our bodies sometimes rebel. Symptoms like regular tummy aches, repeated bouts of diarrhoea, or severe fatigue do not occur for no reason, and nobody should be expected to just accept them as 'one of those things'. Even if your doctor has ruled out coeliac disease or congenital fructose intolerance, nobody can deny you the right to avoid certain foods if you notice that doing so improves your general wellbeing.

Apart from this general overconsumption, antibiotics, high stress levels, or gastrointestinal infections, for example, can also trigger temporary sensitivities to certain foods. When the body has returned to a healthy equilibrium, even a sensitive gut can usually sort itself out. Then there is no need to impose a lifelong ban on certain products, but simply to make sure you consume them in quantities that your system can easily cope with.

A Few Facts About Faeces

Components

Colour

Consistency

Gentle reader, it is now time to get down to the nitty-gritty. So, tighten your trouser braces, push your glasses up the bridge of your nose, and take a good gulp of your tea! While maintaining a safe distance, we must now take a closer look at the mysteries of number twos!

COMPONENTS

Many people believe that faeces are made up mainly of what they have eaten. This is not the case.

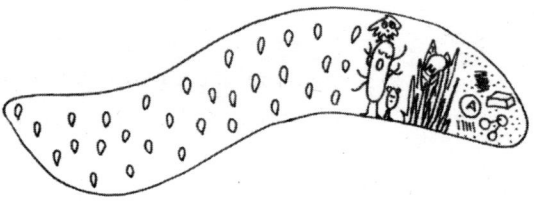

Faeces are three-quarters water. We lose around 100 millilitres of fluid in this way per day. During a passage through our digestive system, some 9.8 litres are re-absorbed. What we deliver into the toilet bowl is the result of an absolute maximum level of efficiency. Whatever fluid is left in it, belongs there. This optimum water content makes our faeces soft enough to ensure that our metabolic waste products can be transported out of our bodies safely.

A third of the solid components are bacteria. They are gut flora that have ended their careers in the digestive business and are ready to retire from the workplace.

Another third is made up of indigestible vegetable fibre. The more fruit and vegetables you eat, the more faeces you excrete per bowel movement. Increasing the proportion of that food group in the diet can raise the weight of a poo from the average 100 to 200 grams, to as much as 500 grams per day.

The remaining third is a mixed bag. It is made up of substances that the body wants to get rid of — such as the remains of medicines, food colorants, or cholesterol.

COLOUR

The natural colour of human faeces ranges from brown to yellowish-brown. Even when we have not eaten anything of this colour. The same is true of our urine — it always tends towards the yellow. This is due to a very important product that we manufacture freshly every day: blood. Our bodies create 2.4 million new blood corpuscles a day. But the same number are broken down every day, too. In that process, the red pigment they contain is first turned green, then yellow — you can see the same process in the various stages of a bruise on your skin. A small portion of this yellow pigment is excreted directly in our urine.

Most of it, though, passes through the liver and into the gut. There, bacteria change its colour once again — this time turning it brown. Examining the colour of faeces can provide a useful insight into the goings-on of our guts:

LIGHT BROWN TO YELLOW: *This colour can be the result of the harmless disorder called Gilbert's syndrome (or Gilbert-Meulengracht syndrome). In this condition, one of the enzymes involved in the breakdown of blood works at only 30 per cent of the normal efficiency. This means less pigment finds its way into the gut. Affecting around 8 per cent of the population, Gilbert's syndrome is relatively common. This enzyme defect is not harmful, causing barely any problems for those who have it. The only side effect is a reduced tolerance for paracetamol, which should be avoided by those with Gilbert's syndrome.*

Another possible cause of yellowish faeces is problems with the bacteria in the gut. If they are not working as they should, the familiar brown pigment will not be produced. Antibiotics or diarrhoea can cause such an alteration in faeces colour.

LIGHT BROWN TO GREY: *If the connection between the liver and the gut is blocked by a kink in the tubes or by pressure (usually behind the gall bladder), no blood pigment can make it into the faeces. Blocked connections are never a good thing, and those who notice a grey tint to their faeces should consult their doctor.*

BLACK OR RED: *Congealed blood is black; fresh blood is red. In this case, the colour is not due to the pigment that can be turned brown by bacteria, but to the presence of entire blood corpuscles. For those with haemorrhoids, a little bright-red blood in the stool is no reason to worry. However, anything darker in colour than fresh, bright-red blood should be checked by a doctor — unless you have been eating large amounts of beetroot.*

CONSISTENCY

The Bristol stool scale was first published in 1997. So it is not very old, when you consider the millions of years that stool has existed. The scale classifies the consistency of faeces into seven groups. A chart like this can be a useful tool, since most people are reluctant to talk about the appearance of their poo. That's perfectly natural. There are some aspects of private life we prefer not to rub other people's noses in! But such a reticence to talk about what we find in the toilet bowl means that people with unhealthy-looking faeces are often unaware of it. They think everybody's business looks like their own. A healthy digestive system, producing faeces with the optimum water content, will produce types 3 or 4. The other types are less than ideal. If they do appear, a good doctor should be able to find out whether your loose stool or constipation is the result of a food intolerance, for example. The chart was developed by Dr Ken Heaton at the University of Bristol in the UK.

The type that a person's faeces belongs to can be an indication of how long indigestible particles take to pass through their gut. According to this, in Type 1, digestive remains take around one hundred hours to pass through the system (constipation). In Type 7, they pass through in just ten hours (diarrhoea). Type 4 is considered the ideal, because it has the optimum ratio between fluid and solid content. Those who find Type 3 or 4 in the toilet bowl may also want to observe how quickly their faeces sink in water. Ideally, they should not plummet straight to the bottom, as this would indicate the possibility that they still contain nutrients that have not been digested properly. Faeces that sink slowly contain bubbles of gas that keep them afloat in water. These are produced by gut bacteria that mostly perform useful services. So this is a good sign, as long as it is not accompanied by flatulence.

Type 1: *Separate hard lumps, like nuts (hard to pass)*

Type 2: *Sausage-shaped, but lumpy*

Type 3: *Like a sausage but with cracks on its surface*

Type 4: *Like a sausage or snake, smooth and soft (note: or like toothpaste)*

Type 5: *Soft blobs with clear-cut edges (passed easily)*

Type 6: *Fluffy pieces with ragged edges, a mushy stool*

Type 7: *Watery, no solid pieces. Entirely liquid*

These have been a few selected facts about faeces, gentle reader. You can now loosen your braces again, and let your glasses slide back down your nose to where they are most comfortable. Here endeth the first chapter of the story of the gut and its goings-on. We now turn to the electricty of life: the nerves.

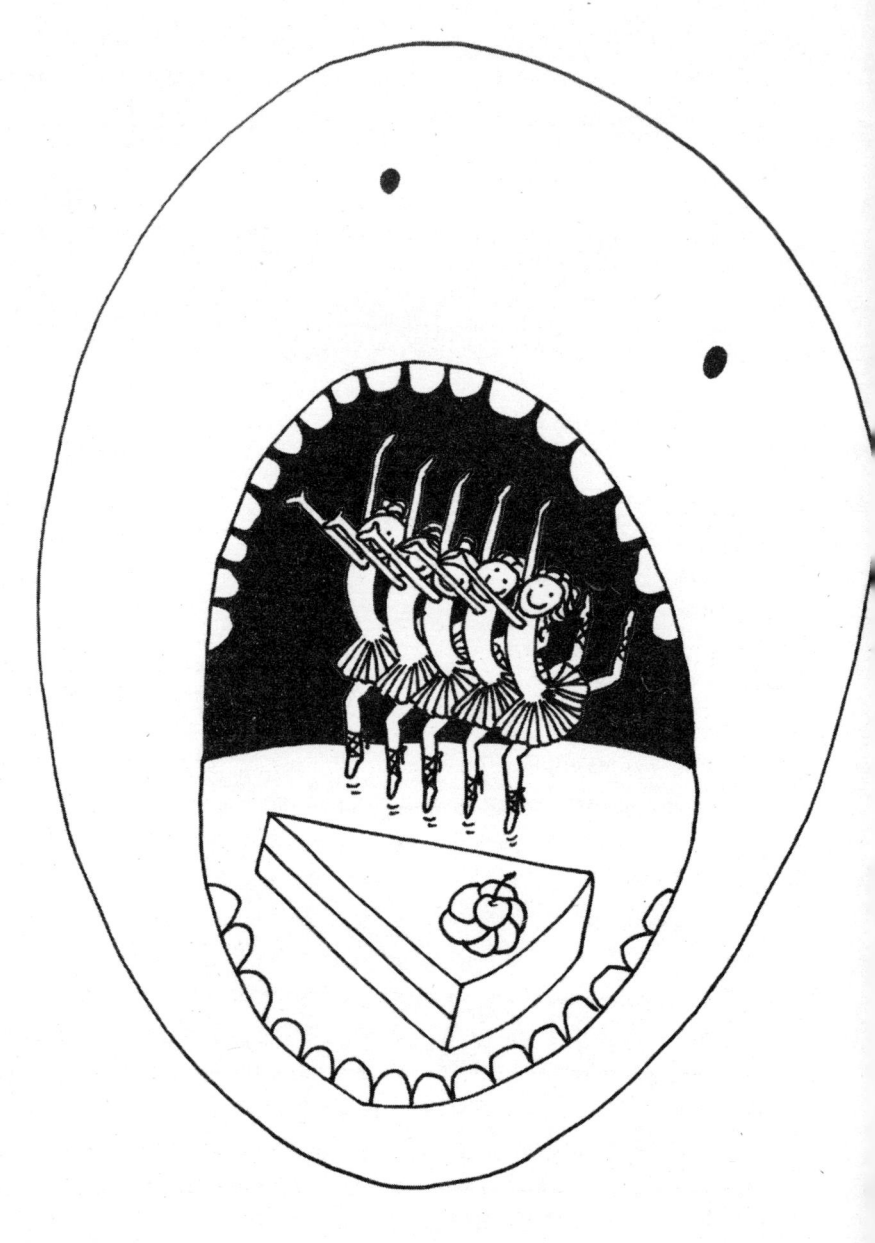

2
THE NERVOUS SYSTEM OF THE GUT

There are places where the unconscious and the conscious meet. When you are sitting at home, eating your lunch, you may be unaware that your next-door neighbour is just metres away, beyond the dividing wall, chomping away on his lunch, too. You might hear the faint creak of his floorboards and suddenly, your awareness reaches out beyond your own four walls. Similarly, there are areas of our own bodies we are simply unaware of. You don't feel your organs working away all day long. When you eat a piece of cake, you taste it while it is still in your mouth, and you are also conscious of the first few centimetres it passes through after you swallow it. But then, as if by magic, the cake is gone! From then on, what we eat disappears into the realm of what scientists call 'smooth muscle'.

Smooth muscle is not under our conscious control. Under the microscope, it looks very different from the tissue of the muscles we can control consciously, such as the biceps. We can flex and relax the muscles in our upper arms at will. Such muscles are made up of the tiniest little fibres, lined up so neatly they look as if they were drawn using a ruler.

The microscopic structure of smooth muscle resembles an organic network, and it moves in mellifluous waves. Our blood vessels are surrounded by smooth-muscle tissue, which explains why we blush when we are embarrassed. Smooth-muscle tissue slackens in response to emotions such as embarrassment, causing the blood capillaries in the skin of the face to dilate. In many people, stress has the opposite effect, causing the muscles surrounding the blood vessels to contract, restricting the flow of blood. This can lead to high blood pressure.

The gut is cosseted by no less than three coats of smooth-muscle tissue. This makes it incredibly supple and able to execute different choreographies in different places. The choreographer directing these muscles is the gut's own ('enteric') nervous system. It controls all processes that take place in the digestive tract, and is extraordinarily autonomous. If the connection between the enteric nervous system and the brain is severed, the digestive tract carries blithely on as if nothing has happened. This property is unique to the enteric nervous system and is found nowhere else in the human body. Without it, our legs would be lame; our lungs would no longer be able to breathe. It is a shame that we are oblivious to the workings of these independent-minded nerve fibres. Burping or breaking wind might sound a bit gross, but the movements involved are as delicate and complex as those of a ballerina.

How Our Organs Transport Food

Allow me to take you on a journey. Let us accompany that piece of cake on its travels in the realm of smooth muscle.

Eyes

Particles of light bouncing off the piece of cake hit the optic nerves at the back of the eyes, generating a nerve impulse. This 'first impression' travels right through the brain to the visual cortex at the back, just below where a high ponytail would be. There, the brain interprets the nerve impulses to form an image — it is not until this happens that we really see the piece of cake. This delicious news is then passed on to the systems that control salivation, with mouth-watering results. Similarly, the mere sight of a yummy treat causes the stomach to produce digestive juices in anticipation.

Nose

If you stick your finger up your nose, you will notice that the cavity continues upward far beyond the reach of your finger. This is where the olfactory nerves are, which are responsible for smelling. They are coated in a protective layer of mucous, so anything we smell must first be dissolved in that slimy substance if it is to get through to the nerves.

Olfactory nerves are specialists — there are specific receptors for a large range of individual smells. Some spend years hanging around up your nose, waiting for their chance to shine. When that single, long-awaited lily-of-the-valley scent molecule finally attaches itself, the receptor proudly calls out 'lily-of-the-valley!' to the brain. Then it might be idle again for the next couple of years. Incidentally, although we are equipped with a large number of olfactory cells, dogs have inconceivably more.

For us to smell it, molecules from the piece of cake first have to drift into the air, to be sucked in through our nostrils as we breathe. They may be aromatic molecules of vanilla, minute plastic molecules from cheap party forks, or evaporating alcohol fragrances from the cake's rum filling. Our olfactory organ is a royal taster with a thorough knowledge of chemistry. The closer we bring the cake-laden fork to our cake-hole, the more detached cake molecules stream into our nose. If we detect tiny traces of alcohol as the cake covers the last few centimetres, we may back our arm up in suspicion, to allow our eyes to inspect the cake again, just to check whether it is supposed to contain alcohol, or whether the fruit in it has started to rot. With all checks passed, it's: mouth open, fork in, and let the ballet begin.

Mouth

The mouth is a place of superlatives. The most powerful muscles in our bodies are the jaw muscles; the body's most flexible striated (not smooth) muscle is the tongue. Working together, they are not only incredible crunchers; they are also nimble manipulators. Another candidate for the record books is tooth enamel — it is the hardest substance produced by the human body. And it needs to be, since our jaws can exert a pressure of up to 80 kilograms on each of our molars — or approximately the weight of a grown man. When we encounter something hard in our food, we pound it with almost the equivalent force of an entire football team jumping up and down on it before we swallow it. All that power is not necessary to deal with a piece of cake — just a few girls in tutus and ballet slippers will suffice.

The tongue plays an important part in mastication. It acts like a football coach, gathering any bits of cake that may be hiding from the chewing action and guiding them back into the game. When the mouthful of cake is sufficiently mushy, it is ready for swallowing. The tongue rounds up about 20 millilitres of cake mush and presses it against the palate, the stage curtain of the oesophagus. It works like a light switch: when the tongue presses against it, the swallowing reflex starts automatically. We close our mouth, since breathing must stop for swallowing to take place. The ball of cake mush — known medically as the 'bolus' — now makes its way towards the pharyngeal area — and it's time for the dancers to enter the stage and for the show to start.

Pharynx

The velum (or soft palate) and the superior pharyngeal constrictor muscle are the two formations responsible for officially closing the connections to the nose. This movement is so powerful that it can be heard down the corridor and around the corner — that popping sound in the ear that often accompanies a powerful swallow. The vocal chords are silenced, and have to be closed. The epiglottis rises majestically, like an orchestra conductor (you can feel it when you place a hand on your neck), the entire base of the mouth is lowered, and a powerful wave pushes the bit of cake into the oesophagus, amid tumultuous applause from the salivary system.

Oesophagus

The cake bolus takes around five to ten seconds to reach this stage. When we swallow, our oesophagus executes a kind of Mexican wave. On the arrival of the bolus, the oesophagus widens to let it pass, closing again behind it. This prevents anything from slipping back up the wrong way.

This process is so automatic that it even works when the owner of the oesophagus is standing on her head. Our piece of cake meanders in this way — oblivious to gravity — gracefully through the upper body. Break dancers would call this move *the snake* or *the worm*; doctors prefer to call it propulsive peristalsis. The top third of the oesophagus is surrounded by striated muscle — that is why we are still aware of the cake passing through that part of the gullet. The unconscious inner world begins at the level of that small hollow you can feel at the top of your breastbone. From there on, the oesophagus is made of smooth muscle.

The oesophagus is sealed at the bottom end by a ring-shaped, sphincter muscle. Taking its cue from the peristaltic motion above, that muscle relaxes for eight frolicsome seconds. This opens the way, allowing the piece of cake to plop unhindered into the stomach. The muscle then closes again, and normal breathing service resumes up in the pharynx.

The journey from mouth to stomach is the first act of the performance. It requires maximum concentration and good teamwork. The conscious, peripheral nervous system and the unconscious, autonomous nervous system must work together in perfect harmony. This cooperation must be well rehearsed. We begin practising swallowing as unborn babies in the womb. We swallow up to half a litre of amniotic fluid a day during this test phase. If something goes wrong at this stage, no harm is done. Since we are completely surrounded by liquid, and our lungs are full of it anyway, we are at no risk of choking in the normal sense.

As adults, we swallow somewhere between six hundred and two thousand times a day. And each act of swallowing involves more than twenty pairs of muscles. Despite this frequency and complexity, things rarely go wrong. In old age, we are more prone to choking. The muscles that coordinate the process may no longer work quite so precisely, the superior pharyngeal constrictor muscle

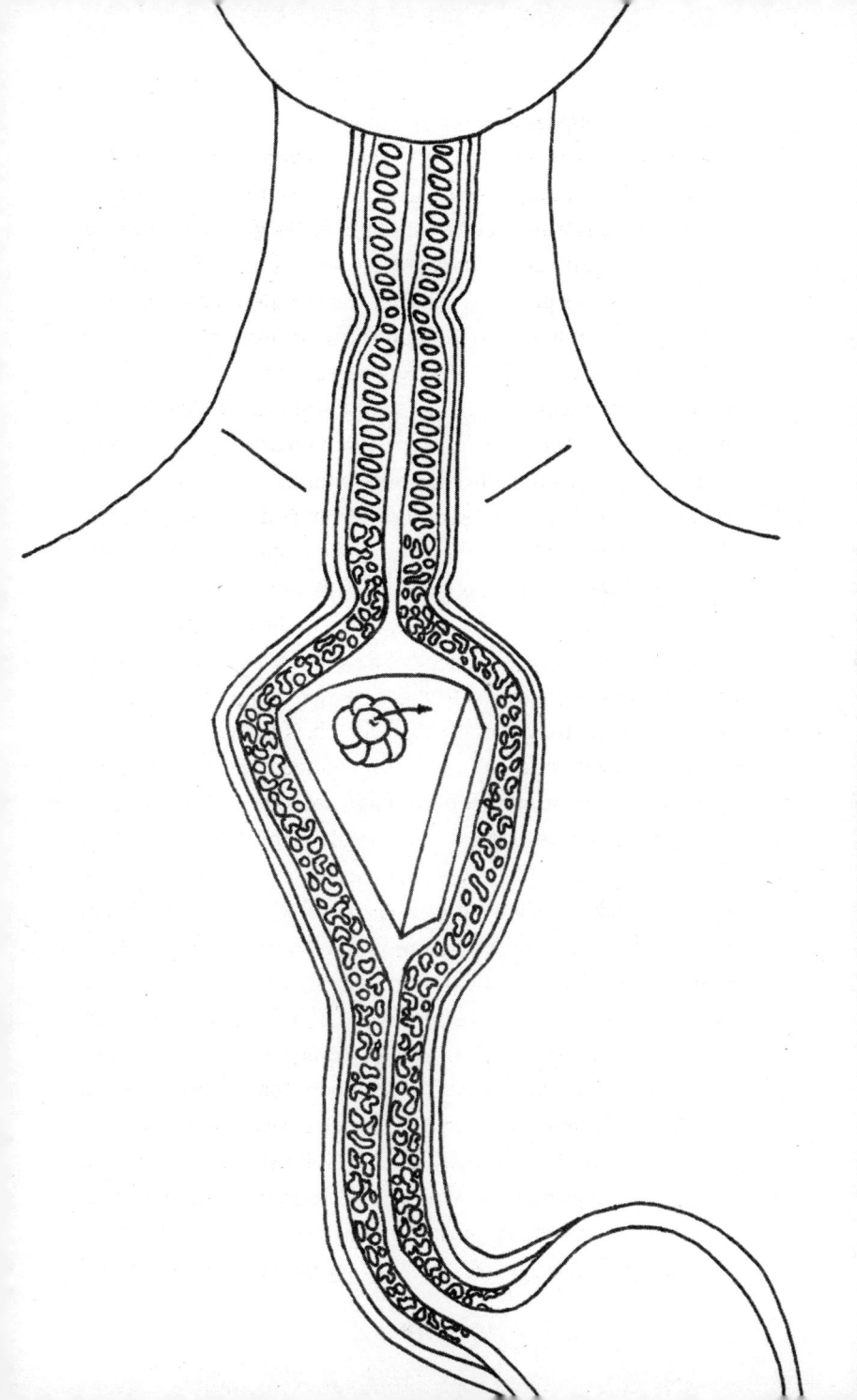

might not be quite the strict time-keeper it was in its youth, or the epiglottal conductor may need the aid of a stick to climb up to the podium. Pounding someone on the back when they are choking is a well-meant gesture — but does little more than startle the ageing pharyngeal team unnecessarily. A better strategy is to seek out a speech therapist to help you whip your swallowing squad into good shape, before choking attacks become too frequent.

Stomach

The stomach is much more of a mover than many people think. Shortly before our piece of cake plops in, it relaxes to accommodate it — and it can keep relaxing and stretching for as long as food keeps arriving. It will make room for as much as we can guzzle. A kilo of cake about the size of a carton of milk will easily fit into this stretchable swing hammock of a stomach. Emotions like fear or stress can reduce the ability of the smooth muscle to stretch, making us feel full, or even nauseous, after eating just a small portion of food.

Once the cake arrives, the walls of the stomach speed up their movements, just like the legs of a person taking a run-up, then — bam! — the food gets a big push. Describing an elegant arc, it is lobbed against the stomach wall, bounces off it, and plops back down. Medics call this process retropulsion; older brothers and sisters call it 'let's see how far I can throw you'. This run-up, push, and plop process is what causes the gurgling sound you can hear if you press your ear against the top of someone's belly (in the little triangle where the ribs meet). When the stomach starts merrily swinging to and fro, the rest of the digestive tract is galvanised into moving, too. This leads the gut to move its contents on down the line, making room for the next batch. That's why we often feel the urge to seek out the toilet soon after enjoying a large meal.

A piece of cake can really get things going in the belly region. The stomach will churn it for about two hours, grinding the mouthfuls into tiny particles, most of them less than 0.2 centimetres big. Scraps of that size are no longer lobbed against the stomach wall, but slip through a little hole at the end of the stomach. This hole is the next sphincter — the doorman who guards the stomach's exit and the entrance to the small intestine.

Simple carbohydrates such as cake sponge, rice, or pasta make it through to the small intestine pretty quickly. There, they are digested and rapidly cause an increase in the levels of sugar in our blood. The doorman detains proteins and fats in the stomach for considerably longer. A piece of steak may easily be churned about for six hours before all of it has disappeared into the small intestine. This explains why we often fancy a sweet dessert after eating meat or fatty, fried foods. Our blood-sugar levels are impatient and want to rise quickly, and dessert provides a quick blood-sugar fix. Meals rich in carbohydrates may perk us up more quickly, but do not keep us feeling full for as long as meaty or fatty meals.

Small intestine

When the mini-morsels reach the small intestine, the real process of digestion begins. As it passes through this tube, the motley cake mush will almost completely disappear into its walls — a bit like Harry Potter on platform 9¾. The small intestine pluckily pounces on the piece of cake. It squeezes it, hashes it up from all sides, wiggles its villi among what we might now call the 'cake chyme', and, when it is thoroughly mixed, moves it on down the digestive line. Under the microscope, we can see that even the microvilli help it along. They move up and down like tiny trampling feet. Everything is in motion.

Whatever our small intestine does, it always obeys one basic rule: onward, ever onward! This is achieved by the peristaltic reflex. The man who first discovered this mechanism did so by isolating a piece of gut and blowing air into it through a small tube — and the friendly gut blew right back. This is why many doctors recommend a high-fibre diet to encourage digestion: indigestible fibre presses against the gut wall, which becomes intrigued and presses back. These gut gymnastics speed up the passage of food through the system and make sure it remains supple.

If our cake chyme were to listen carefully, it might hear a 'heave!' The wall of the small intestine contains a particularly large number of pacemaker cells, which emit tiny bioelectric pulses. For the muscles of the gut, this is as if someone were to shout 'heave!', and then again, 'heave!' In this way, the muscle is prevented from drifting off course, and 'heaves' back into place, like a clubber on the dance floor responding to the beat. This keeps the piece of cake, or what's left of it, moving unerringly onward.

The small intestine is the hardest-working part of our digestive tract, and is very diligent about doing its job. There is only one unequivocal exceptional case when it does not see a digestive project through to the end: when we throw up. The small intestine is quite pragmatic when we need to vomit. It does not invest work in something that will not do us good. It simply sends such stuff straight back by return of post.

Apart from a few remnants, our piece of cake has now disappeared entirely into the bloodstream. We could now follow those stragglers as they pass into the large intestine — but then we would miss a mysterious and oft-misunderstood creature that we can hear but not see. So let's stay a little while longer and get acquainted with it.

After digestion, only a few rough leftovers remain in the stomach and small intestine — such as an unchewed maize kernel, tablets coated to stop them dissolving in gastric juices, surviving bacteria from the food we have eaten, or a piece of chewing gum swallowed by mistake, for example. Now, the small intestine is a stickler for cleanliness. It is one of those types who cleans up the kitchen straight after a meal. If you were to pay your small intestine a visit just two hours after it has finished digesting something, you would find everything spick and span, with barely a whiff of what went on there a short time ago.

An hour after the small intestine has digested a food item, it begins the cleaning-up process. The scientific name for this process is 'migrating motor complex'. When it kicks in, the stomach doorman is kind enough to open the gates again to allow these leftovers to be herded into the small intestine. It then moves them along with a wave powerful enough to sweep everything along with it. When observed with a camera, this looks so cute that even sober-minded scientists can't help but nickname the migrating motor complex the little 'housekeeper'.

Everyone has heard their little housekeeper at work. It is the rumbling tummy, which, contrary to popular belief, does not come mainly from the stomach, but from the small intestine. Our tummies don't rumble when we're hungry, but when there is a long-enough break between meals to finally get some cleaning done. When the stomach and the small intestine are both empty, the coast is clear for the housekeeper to do its work. If the stomach is involved in the lengthy process of grinding down a steak, the housekeeper just has to be patient — only after six hours of being churned in the stomach and around five hours of being digested in the small intestine is the steak safely gone, after which the housekeeper can start clearing up. We don't necessarily always hear the housekeeper at work — it depends on how much air has found its way into the stomach and the small intestine. If we eat something before the clean-up is finished, the housekeeper immediately stops working and returns to waiting mode. Food needs to be digested in peace, and not swept ahead too soon in a cleaning frenzy. Constant snacking means there is no time for cleaning. This is part of the reason why some nutritional scientists recommend we leave five hours between meals, although there is no scientific evidence proving that the interval must be precisely five hours. Those who chew their food thoroughly create less work for their housekeeper and can listen to their tummy when it tells them it's time to eat again.

Large intestine

At the end of the small intestine is a structure known as Bauhin's valve. It separates the small from the large intestine, which is good because the two neighbours have very different ideas about work ethics. The large intestine is a much more leisurely type. Its motto is not really 'onward, ever onward'; it is not averse to shifting what remains of our food backwards as well as forwards — if it feels right at the time. It has no migrating housekeeper. The large intestine is the tranquil home of our gut flora, which deal with anything that gets swept into the large intestine undigested.

The large intestine works at a more leisurely pace because it has to consider several different players. Our brain is picky about when it wants us to go to the toilet; the bacteria in our gut wants ample time to deal with undigested food; and the rest of our body very much wants to get back the fluids it lent to the digestive system.

What makes it as far as the large intestine no longer resembles a piece of cake — and nor should it. The unabsorbed remains of the cake may include a few fruit fibres from the cherry on the top, and the rest is made up of digestive juices that are reabsorbed here. When we are anxious, our brain jockeys the large intestine along, leaving it without sufficient time to reabsorb all that fluid. The result is diarrhoea.

Although the large intestine (like the small intestine) is a smooth tube, it is always shown in diagrams looking like a lumpy string of beads. Why is that? In fact, that is what the large intestine looks like when the abdomen is opened up. The simple reason for this is that it is engaged in a slow-motion dance. Just like the small intestine, it bulges as it processes the food it receives, so as to hold it where it needs it to be. However, it tends to remain in one position for a long time without moving — something like those street mimes who stand in one statuesque pose until someone comes along and puts a coin in their hat. Every now and then, it relaxes and forms bulges in other places, and then remains in that configuration for a while. Anatomy books depict it in this way — like a child who blinks as the class photo is taken, and thereby will appear in the yearbook looking dopey for evermore.

Two or three times a day, the large intestine stirs from its slumbers and gives an enthusiastic shove to the concentrated food mush to push it forward. Those who provide their large intestine with sufficient bulk may even have to go to the toilet two or three times a day. For most people, the content of their large intestine is enough for one bowel movement a day. But, statistically speaking, three times a day is still a healthy frequency. Women's large intestines are generally slightly more lethargic than men's. Medical researchers have not yet discovered why this is so — but the greatest likelihood is that it has a hormonal cause.

The cake's journey from fork to toilet takes one day on average. Faster guts accomplish it in eight hours; slower digesters can take three-and-a-half days. Due to all the mixing they undergo, some cake particles may linger in the chill-out space of the large intestine for twelve hours; others might lounge around for up to 42 hours. Leisurely digesters should not worry, as long as the consistency is fine and they have no other complaints. On the contrary, one Dutch study showed that those who belong to the 'once a day or less' faction and those who have occasional constipation are less likely to contract certain rectal diseases — consistent with the motto of the large intestine: 'slow and steady wins the race'.

Reflux

The stomach can sometimes stumble. Its smooth-muscle tissue can trip up just like the striated muscles of the legs. When that causes a substance like gastric acid to reach places that are not designed to deal with it, it hurts. Reflux is the regurgitation of gastric acid and digestive enzymes into the pharyngeal area. In the case of heartburn, those juices come into contact with the end of the oesophagus, causing a burning sensation in the chest.

Reflux has the same cause as stumbling: it's all down to the nerves. They control our muscles. If the nerves in our eyes fail to detect a doorstep in time, our legs receive incorrect information, and carry on walking as if there were no obstacle in the way — and we stumble. When the nerves of our digestive system receive incorrect information, they fail to keep our gastric juices where they belong, and allow them to start moving in the wrong direction.

The junction between the oesophagus and the stomach is an area that is particularly susceptible to such stumbles. Despite safety measures that include a narrow oesophagus, a steadied position in the diaphragm, and the curve at the entrance to the stomach, things sometimes still go wrong. Around a quarter of the German population regularly experience such problems. This is not just another modern fad: nomadic people, whose way of life has not changed for hundreds of years, have similar rates of heartburn and reflux as Germans.

The crux of the matter is that two different nervous systems have to work together in the oesophagus and stomach area — the nervous system of the brain, and that of the gut. For example, the sphincter muscle between the oesophagus and the stomach is under the control of nerves from the brain. The brain also influences gastric-acid production. The nerves of the digestive tract ensure that the oesophagus moves things downwards in a harmonious Mexican wave, keeping it clean with the thousand-or-so times we swallow saliva over the course of a day.

Practical tips to help with heartburn and reflux are based on trying to get those two nervous systems back on the right path. Chewing gum or sipping tea can help the digestive tract because small, repeated swallows help nudge the nerves in the right direction — down towards the stomach, not back up. Relaxation techniques can help persuade the brain not to send out such hectic instructions via the nerves. In a best-case scenario, that should result in constant closure of the sphincter, leading to less acid production.

Cigarette smoke stimulates areas of the brain that are also activated by eating. This may lead to a sense of satisfaction, but it also tricks the brain into producing more gastric acid for no practical reason, as well as causing the sphincter between the oesophagus and stomach to relax. This is why giving up smoking often helps reduce reflux and heartburn complaints.

Pregnancy hormones can cause similar disarray. Their intended function is to keep the womb relaxed and cosy for the unborn baby. But they also have a similar effect on the sphincter of the oesophagus. The result is a leaky connection to the stomach, which, combined with the pressure of the bulging tummy from below, causes acid to rise. Contraceptives containing female hormones can cause reflux as a side effect.

Cigarette smoke or pregnancy hormones — our nerves are not like fully insulated electric cables. They are embedded organically in our tissues, and react to the substances around them. That's why many doctors recommend avoiding foodstuffs that can reduce the strength of the sphincter that seals the stomach off from the oesophagus: chocolate, hot spices, alcohol, sugary sweets, coffee, and so on.

All these substances influence our nerves, but they do not necessarily cause an acid-stomach stumble in everybody. The results of American studies indicate that each individual should use trial and error to find out which foods affect their nerves, rather than unnecessarily cutting out everything that might be the culprit.

There is an interesting connection discovered by means of a drug that was never approved because of its side effects. This substance blocks nerves at a location where glutamate normally binds with them. Most people are familiar with glutamate as a flavour enhancer, but it is also released by our nerves. In the nerves of the tongue, glutamate causes an intensification of taste signals. This can create confusion in the stomach, since the nerves do not know whether the glutamate originates from their colleagues elsewhere in the body, or from the local Chinese takeaway. The trial-and-error test here would involve avoiding food rich in glutamate for a while. To do this, you will have to take your reading glasses to the supermarket to check those lists of ingredients in tiny print on food labels. Often, glutamate hides behind more complicated formulations such as 'monosodium glutamate' or similar. If avoiding it results in an improvement — fine. If not, at least you will have eaten more healthily for a while.

Those whose stomach stumbles less than once a week can turn to simple remedies for relief, such as antacids from the pharmacy, or household cures like raw potato juice. Neutralising stomach acid should not be used as a long-term strategy, however. Stomach acid is useful for combating the harmful effects of allergens and bacteria from our food, and is instrumental in digesting proteins. What is more, some antacid medications contain aluminium, which is a very alien substance for our body to deal with. So, overuse of antacids should be avoided. Always read and follow the instructions on the leaflet.

Reliance on antacids for four weeks or more should be seen as a warning sign. Anyone who ignores such warnings will soon feel the wrath of a disgruntled stomach that wants its acid back — to make up for the effects of the medication for one thing, and also to return to its natural acidic state. Antacids are never a long-term solution, and certainly not for other acidic phenomena, such as gastritis — the medical name for an inflammation of the lining of the stomach.

When symptoms continue despite the use of antacids, doctors must get creative to find out why. If blood tests and a thorough physical examination yield no abnormal results, a doctor may prescribe a drug called a proton pump inhibitor, or PPI. PPIs inhibit the production of acid and its secretion into the stomach. Short-term use of a PPI may leave the stomach lacking a little acid; but, for those who suffer from heartburn or reflux, they provide a short respite for the stomach and the oesophagus so that they can recover from the effects of such acid attacks.

If attacks strike mainly at night, it is a good idea to try propping the upper body up to an angle of 30°. This can involve complex pillow constructions and the use of a protractor at bedtime, but specially manufactured pillows are also available from specialist suppliers. Incidentally, a 30° upper-body inclination is also good for the cardiovascular system. My old physiology professor never tired of telling us this, and, given that his speciality is cardiovascular research, I am more than inclined to believe him. But it also means I now can't help imagining him propped up in bed every time someone mentions his name.

Patients *should* lose sleep over such serious warning signs as difficulty in swallowing, weight loss, swelling, or any sign of blood. When such symptoms appear, it is high time for an exploratory expedition into the stomach with a camera — no matter how unpleasant an experience that can be. The real danger with reflux does not, in fact, come from burning acid, but from bile reaching the oesophagus from the small intestine via the stomach. Bile does not cause a burning sensation, but has much more insidious effects than acid. Luckily, for most people who suffer from reflux, there is very little bile acid involved.

The presence of bile acid can seriously confuse the cells of the oesophagus. Suddenly, they are no longer sure where they are, thinking 'Am I really an oesophagus cell? I keep sensing bile! Perhaps I've really been a small intestine cell all these years without realising it ... Silly me!' Anxious to do the right thing, they change from oesophagus cells into gastrointestinal cells. That can cause problems. Mutating cells can make mistakes in their own programming, and no longer grow in a controlled way like other cells. However, this has serious repercussions for only a small percentage of people who suffer from stomach 'stumbles'.

In the vast majority of cases, reflux and heartburn are 'stumbles' which are unpleasant, but not dangerous. When we stumble while walking, we usually briefly adjust our clothing, neutralise the shock with a shake of the head, and walk on at a measured pace. The same reaction is appropriate in the case of an acid stumble of the stomach — take a couple of mouthfuls of water for adjustment, neutralise the acid if necessary, and then continue walking, perhaps at a less hectic pace.

Vomiting

If you took a hundred people on the verge of vomiting, you would have a very mixed group. Person number 14 is on a rollercoaster, hands high in the air; person 32 is cursing the egg salad she ate; number 77 is holding a pregnancy test in disbelief; and number 100 has just read the words 'may cause nausea and vomiting' in the leaflet that came with his new medicine.

Vomiting is not a stomach stumble: it happens according to a precise plan. It is a tour de force performance. Millions of tiny receptors test our stomach contents, examine our blood, and process impressions from the brain. Every piece of information is gathered in the huge fibrous network of our nervous system and sent to the brain. The brain evaluates this information. Depending on how many alarm bells are ringing, it makes a decision: to barf or not to barf. The brain transmits its decision to selected muscle groups, and they get down to work.

If you were to X-ray the same hundred people while they were vomiting, the picture would be the same a hundred times over. The brain responds to the alarm, activates the area responsible for vomiting, and switches the body to emergency mode. We turn pale as the blood drains from our cheeks and is sent to the abdomen. Our blood pressure drops, and our heart rate falls. Finally, we feel that unmistakable sign: saliva, and lots of it. The mouth begins producing it in great quantities as soon as it receives information from the brain about the emergency that's underway. This saliva is meant to protect our teeth from the corrosive effects of the gastric acid they are about to come into contact with.

To begin with, the stomach and gut move in small, nervous

waves — shoving their contents in completely opposite directions as they begin to panic slightly. We cannot feel this toing-and-froing, as it takes place in the realm of smooth muscle. However, this is about the time that most people realise intuitively that they should seek out a suitable receptacle.

An empty stomach is no defence against vomiting, since the small intestine is just as able to expel its contents. For this to happen, the stomach opens the gate to allow the contents of the small intestine back in — every member of the team works together to bring this major project to completion. When the small intestine suddenly sweeps its contents into the stomach, sensitive nerves there are stimulated. These nerves respond by sending signals to the vomit-control centre in the brain. Now there's no doubt about it: everything is ready for the big heave.

Our lungs take a particularly large breath before our airways are closed. The stomach and the opening to the oesophagus suddenly relax and — bam! — the diaphragm and abdominal muscles abruptly press upwards, squeezing us like a tube of toothpaste. The entire contents of the stomach are then propelled from the body like a pilot in an ejector seat.

Why we vomit and what we can do to prevent it

The human animal is especially designed to be able to vomit. Other animals with this ability include apes, dogs, cats, pigs, fish, and birds. Those that are not able to vomit include mice, rats, guinea pigs, rabbits, and horses. Their oesophagus is too long and narrow, and they lack the nerves that are so talented at vomiting.

Animals that cannot vomit have to have different eating habits to ours. Rats and mice nibble at their food, biting off tiny pieces to test their suitability. They only continue to eat when they are sure the trial nibble has not done them any harm. If it turns out to be toxic, the most they suffer will be a bout of tummy ache. They also learn not to try to eat it again. Furthermore, rodents are much better than us at breaking down toxins because their liver has more of the necessary enzymes. Horses, however, are not even able to nibble. If something bad ends up in their small intestine, the results can often be life threatening. So, really, we have reason to be proud of our body's abilities whenever we find ourselves crouched over the toilet bowl, 'throwing our guts up'.

The short rests between retches could be utilised for a little reflection. Person number 32's notorious egg salad seems to have held its form surprisingly well when it returns from its short sojourn in the realm of the stomach. A few pieces of egg, the odd pea, or piece of pasta are still recognisable. It might cross 32's mind that she can't have chewed it very well. Moments later, another retch produces a rather more deconstructed arrangement. Vomit that contains recognisable bits of food is almost certain to have originated from the stomach, and not from the small intestine. The smaller the particles, the more bitter the taste and the more yellow the colour, the more likely it is to be a salutation from the small intestine. Clearly identifiable food may not have been chewed properly, but at least it has been ejected from the stomach quickly, before making it as far as the small intestine.

The way we vomit also tells us quite a bit. Sudden vomiting that comes in a violent surge almost without warning is likely to be caused by a gastrointestinal virus. This is due to the fact that the sensors count how many pathogens they encounter; when they decide the numbers have got out of hand, they slam on the emergency brakes. Below this threshold, the body's immune system could likely have dealt with the situation, but now the job is handed over to the gastrointestinal muscles.

Food or alcohol poisoning also cause vomiting in surges. However, this time we usually get a fair warning beforehand, in the form of nausea. The feeling of nausea is the body's way of telling us

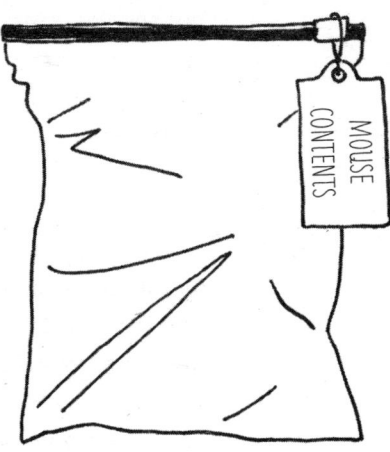

that the food we have eaten is not good for us. Person 32 is likely to be much more wary of that bowl of egg salad on the buffet in future.

Person 14 on the rollercoaster feels just as nauseous as number 32 of egg-salad fame. Rollercoaster puking is basically the same as travel sickness. In this case, no toxins are involved, yet sick still ends up on people's shoes, in the glove compartment, or splattered on the windscreen of the car downwind of yours. The brain is the bodyguard of the body — guarding it meticulously and cautiously, especially when the body belongs to a small child. Currently, the best explanation of motion sickness is this: when the information sent to the brain from the eyes is at odds with that sent by the ears, the brain cannot understand what is going on, and slams on every emergency brake at its disposal.

When a passenger reads a book in a moving car or train, their eyes register 'hardly any motion', while the balance sensors in the ears say 'lots of motion'. It's the same, but opposite, effect as when you watch the trees whizz by when driving through a forest. If you move your head a little as well, it looks as if the trees are rushing by faster than you are actually moving — and that, too, confuses the brain. On an evolutionary scale, our brains are familiar with such mismatches between eyes and balance sensors as signs of poisoning. Anyone who has ever drunk too much or taken drugs will have felt the room spinning, even when they are not moving at all.

Vomiting can also be caused by intense feelings such as emotional strain, stress, or anxiety. Under normal circumstances, we synthesise the stress response hormone CRF (*corticotropin releasing factor*) in the morning, creating a supply to help face the challenges of the day. CRF helps us tap into energy reserves, prevents the immune system from overreacting, and helps our skin tan as a protective response to stress from sunlight. The brain can also inject an extra portion of CRF into the bloodstream if we find ourselves in a particularly upsetting situation.

However, CRF is synthesised not only by brain cells, but also by gastrointestinal cells. Here, too, the signal is: stress and threat! When gastrointestinal cells register large amounts of CRF, irrespective of where they originate (in the brain or in the gut), the information that one of the two is overwhelmed by the outside world alone is enough for the body to react with diarrhoea, nausea, or vomiting.

When the brain is stressed, vomiting expels partly digested food in order to save the energy required to complete the digestive process. The brain can then use that energy to solve the problems at hand. When the gut is stressed, partly digested food is ejected either because it is toxic or because the gut is currently not in a position to digest it properly. In both cases, it can make good sense to press the eject button. There is simply no time for gentle, comfortable digestion. When people throw up from nerves, it is simply their digestive tract trying to do its best to help.

Incidentally, petrels use vomiting as a defence strategy. Vomiting is a sign from the plucky little birds to steer well clear of their nests. Researchers use that to their advantage. They approach a petrel's nest holding out a sick bag, and the seabird pukes right into it. Back at the lab, they can test the petrel vomit for anything, from the presence of heavy metals to the variety of fish it contains. This gives them a measure of how healthy the environment is.

The following includes some simple strategies for reducing unnecessary attacks of vomiting:

1. For travel sickness: keep your eyes fixed on the horizon far ahead. This helps the eyes and balance sensors coordinate their information better.
2. Listen to music on headphones, lie on your side, or try relaxation techniques — some people find this helpful.

One possible explanation for this is that all these activities are generally calming. The more secure we feel, the less we encourage the brain's state of alarm.
3. Ginger. There are now quite a few studies proving that ginger has a beneficial effect. Substances contained in root ginger block the vomit centre of the brain, and the feeling of nausea along with it. Many ginger sweets and candies, however, contain only ginger flavouring, so make sure anything you take contains the genuine stuff.
4. Drugs you can buy from the pharmacy to prevent vomiting work in various ways. They can block the receptors in the vomit centre (the same effect as ginger), numb the nerves of the stomach and gut, or suppress certain alarm signals. The last group are almost identical to drugs used to treat allergies. Both suppress the alarm-signal transmitter histamine. However, the drugs used to prevent vomiting have a much stronger effect on the brain. Modern allergy drugs have been developed and improved to such a degree that they barely dock in the brain at all. This interaction with the brain is what makes the suppression of histamines cause drowsiness.
5. P6! This is an acupuncture point, which is now recognised by Western medicine as effective against nausea and vomiting. Its benefits have been proven in more than forty studies, including placebo-controlled trials. Doctors do not know how or why P6 works. The point is located two to three finger-breadths below the wrist, right between the two prominent tendons of the lower arm. If you don't happen to have an acupuncture needle handy, you can try gently stroking the skin at that point, until symptoms improve. This technique has not been proven in scientific studies, but it may worth trying a little self-experimentation. In traditional Chinese medicine, stimulating this point is believed to activate the energy pathway, or meridian, running up the arm and through the heart, which

relaxes the diaphragm and then runs on through stomach and into the pelvis.

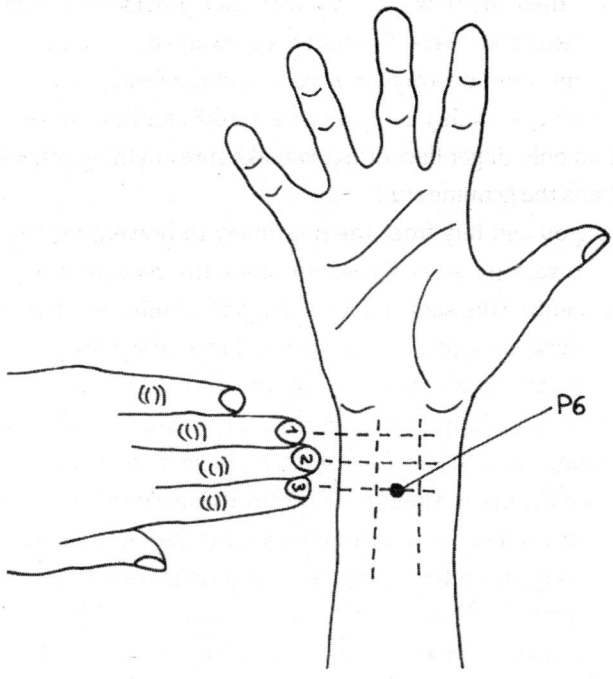

Not every strategy will work for all causes of nausea. Remedies such as ginger, pharmacy-bought medicines, or P6 can help, but for vomiting caused by emotional factors, the best thing is often to build a safe nest for your own inner petrel, so to speak. Relaxation techniques or hypnotherapy (from a reputable practitioner) can help train the nerves to be more thick-skinned. The more often and the longer you practise, the better you will get at it. Silly stress at the office or exam-related anxiety become less threatening when we refuse to let them affect us so personally.

Vomiting is never a punishment from the stomach. Rather, it is a sign that our brain and our gut are willing to sacrifice themselves to the ultimate extreme for us. They protect us from unseen toxins in our food, are over-careful when faced with travel-related eye/ear hallucinations, and save energy to deal with imminent problems. Nausea is meant to give us an orientation for the future, letting us know what is good for us, and what is bad.

If you are unsure what the cause of your nausea is, it is a good idea to simply trust your gut feeling. The same principle applies when you have eaten something bad, but don't feel the need to vomit. You should not force the situation — with a finger down your throat, by drinking salt water, or by having your stomach pumped. Taking acidic or foaming chemicals can cause more problems than it solves. Foam can easily migrate from the stomach to the lungs, and acidic drugs may simply burn the oesophagus for a second time. For these reasons, induced vomiting is no longer a technique normally recommended for use in modern emergency medicine.

True nausea is a programme that has evolved over many millennia, which has the ability to wrest the reins of control from our conscious mind. The conscious mind often reacts to this drastic takeover with shock and indignation — it was blithely planning to order another round of tequila shots, and now, this? But, since it is often the conscious mind that got the body into this vomitous situation in the first place, it eventually has to back down. If vomiting is caused by an unnecessary, overcautious reaction, the conscious mind can always return to the negotiating table and play its anti-vomiting aces.

Constipation

Constipation is like　　　　　. You wait for something that just won't　　　　. And, still you have to use a lot of force. Sometimes, in return for all that effort, you get no more than •••. Or it works, but not very

often.

Between 10 and 20 per cent of people in Germany are constipated. If you want to join this club, you must fulfil at least one of the following conditions: perform a bowel movement less than three times a week, particularly in hard-stool form a quarter of the time, often in pellet form (•••), which is difficult or impossible to pass without help (medication or tricks); and experience no satisfying feeling of emptiness on leaving the toilet.

Constipation results from a disconnect between the nerves and the muscles of the gut, when they are no longer working towards quite the same goal. In most cases, digestion and transportation of food through the system are still working at normal speed — it is not until the very end of the large intestine that disagreement arises as to whether the contents need to be expelled right away or not.

The best parameter for assessing constipation is not *how often* you need to go to the toilet, but *how difficult* it is. Time spent sitting on the toilet is supposed to be a time of splendid isolation and relaxation; but for those whose experience is not so laid-back, it can be a troubling time. There are various levels of constipation: temporary constipation can be due to travelling, illness, or periods of stress, while more obstinate constipation can indicate more long-term problems.

Almost half of us have experienced constipation when travelling. Particularly in the first few days of a trip, it is often hard to go properly. This can be due to a variety of reasons, but in most cases it boils down to the simple fact that the gut is a creature of habit. The nerves of the gut remember what kind of food we prefer, and at what time we prefer to eat it. They know how much we move around and how much water we drink. They know whether it is day or night, and what time we usually go to the toilet. If everything goes according to plan, they complete their tasks without complaint, and activate our gut muscles to help us digest.

When we travel, we have a lot on our minds, such as trying to remember whether we picked up our keys and turned off the iron. We might remember to take a book or some music to keep our brains happy. But there is one thing we always forget: that creature of habit, the gut, is also travelling with us, and it has been suddenly torn from its familiar routine.

We spend the whole day eating pre-packed sandwiches, strange plane meals, or unfamiliar spices. At the time that we would normally be enjoying our lunch break, we're stuck in traffic or waiting at the check-in counter. We drink less than normal, for fear of having to go to the toilet too often, and we dehydrate even further during the flight. And, as if that weren't enough, we might also have to face a big, fat bout of jetlag.

All this does not go unnoticed by the nerves of the gut. They can get confused and put the brakes on until they receive a signal that everything is normal and they can start work again. Even when the gut has done its work despite the confusion, and signals to us that we should seek out the toilet, we add to its woes by suppressing the urge because it happens not to be a convenient time. Also, if we're honest, travel constipation can often be caused by 'not my loo' syndrome. Sufferers of this syndrome simply dislike doing their business in unfamiliar toilets. Their biggest challenge is posed by public conveniences. Many people use them only when it's absolutely necessary, construct elaborate 'seat sculptures' out of toilet paper, or crouch what feels like ten kilometres away from the toilet bowl. But even all that doesn't help those with serious cases of 'not my loo' syndrome. They simply cannot relax enough to finish the work that their creature of habit has begun. When this happens, a holiday or business trip can become a rather unpleasant experience.

There are some little tricks that can be useful for people with brief or mild instances of constipation. These tricks can lower inhibitions and help get things moving in the bowel department:

1. There is a certain foodstuff we can eat to nudge the gut wall into action: fibre. Dietary fibre is not digested in the small intestine, and can knock on the wall of the large intestine in a friendly way, to say there is someone here who wants to be shown the way out. The best results are produced by psyllium seed husks and the rather more pleasant-tasting plum. Both contain not only fibre, but also agents that draw extra fluids into the gut — making the whole business smoother. It can take two to three days before their effect is felt, so you can start eating them either a day before your trip or on the first day — whatever feels safer. Those with no plum compartment in their suitcase can buy dietary fibre in tablet or powdered form from their pharmacy or drugstore. Thirty grams is an appropriate daily dose of dietary fibre.

2. There are two kinds of fibre: water-soluble and insoluble. The latter is better at stimulating movement through the digestive system, but can often cause tummy aches. Water-soluble fibre does not provide quite such a powerful push, but it does make the contents of the gut softer and easier to deal with. Nature's design is rather clever: the skins of many fruits contain large amounts of insoluble fibre, while the flesh of the fruit contains more soluble fibre.
3. Consuming dietary fibre is little help if you do not also consume sufficient fluids. Without the presence of water, fibre binds together in solid lumps. Water makes them swell up into balls. This gives the bored gut something to do while your brain enjoys the in-flight entertainment.
4. Drinking more fluids can only help those who don't already drink enough. For those who do, drinking even more will not bring about any improvement. But it is a different story if the body gets too little fluid. The gut reacts by extracting more water from the food passing through it. That makes faeces harder. Small children running high temperatures often lose so much body fluid through sweating that their digestive system grinds to a halt. Air travel can cause the body to lose similar quantities of water, even without sweating. The air in the plane is so dry that it extracts fluid from our bodies without our even noticing. The first sign we have of it is an unusually dry nose. During air travel it is a good idea to try to drink more than normal, to keep the water in your body at a normal level.
5. Don't put yourself under pressure. If you need to go to the toilet, just go — especially if you are a creature of habit like your gut, and usually go at an appointed time. If you normally go to the toilet in the morning, but suppress the urge because you're travelling, it is as if you have broken an unspoken agreement with your gut. Your gut likes to work according to plan. Pushing digested food back into the holding pattern even just a couple

of times trains the nerves and muscles to operate in reverse gear. That can make it increasingly difficult to change gears back again. This is compounded by the fact that the longer the faeces stay in the gut, the more time the body has to extract fluid from them, making the business ever harder. A couple of days of suppressing the urge can lead to constipation. So, if you still have another week of your camping holiday to go, you'd better get over your fear of the communal toilets before it's too late!

6. Probiotics and prebiotics — living, beneficial bacteria and their favourite food can breathe new life into a tired gut. It is a good idea to consult your pharmacist about this, or turn to the relevant section later on in this book.

7. Take more walks? This is not always a successful strategy. A sudden decrease in exercise can cause the gut to slow down, it's true. But for those who already exercise enough, more movement will not help them attain digestive nirvana. Tests have shown that it takes extremely strenuous exercise to achieve a measurable effect on the movement of the gut. So, unless you are planning to engage in some kind of power sports, forcing yourself to take an extra walk will have little effect — on your ability to go to the toilet successfully, at any rate.

Those with a taste for the unusual might want to try the rocking squat technique: sitting on the toilet, bend your upper body forward as far as possible towards your thighs, then straighten up to the sitting position again. Repeat this a few times, and it should begin to work. No one watches you while you are on the toilet, and you have a moment of free time, so what could be a better opportunity for an unusual experiment?

When household remedies and rocking on the toilet fail:

In more stubborn cases of constipation, the nerves of the gut are not just confused or sulking; they are in need of a bit more support from their owner. If you have tried all the little tips and still don't leave the toilet singing a merry tune, it may be time to rummage in another box of tricks. But you should only do this if you already know the reason for your problem. If you don't know the precise cause of your constipation, you cannot choose the right remedy.

If constipation comes on very suddenly, or lasts for an unusually long time, you must consult your doctor. The problems may stem from undiagnosed diabetes or thyroid problems, or you may just be a natural-born slow transporter.

Laxatives

The aim of taking laxatives is easily stated: to produce the perfect little pile. Laxatives can coax even the shyest of guts out of their shell. Laxatives come in various types that work in different ways. For all hopelessly constipated holidaymakers, slow transporters, campsite toilet objectors, or haemorrhoid-hindrance conquerors, here's a look at that box of tricks.

The perfect little pile by means of osmosis

... is well-formed and not too hard. Osmosis is water's sense of equality. When one region of water contains more salt, sugar, or similar, than another, the less rich water will flow towards the richer water, until both contain the same amount of solute and they can live on in peaceful equilibrium. The same principle helps to revive wilting lettuce — simply soak the sad salad in water, and half an hour later your greens will be crispy again. Water flows into the lettuce because its cells contain more salts, sugars, etc. than the pure water in the bowl.

Osmotic laxatives make use of this 'sense of justice'. They contain certain salts, sugars, or tiny molecular chains, which eventually reach the large intestine. As they make their way there, they pick up all the water they can, making the next trip to the toilet a much smoother affair. But overdoing it with such laxatives causes them to extract too much water. Diarrhoea is the sure sign that you have taken too many.

Osmotic laxatives come in two types: you can choose whether to take sugars or salts, or short molecular chains, to help retain water in the gut. The salts, such as sodium sulphate (also known as Glauber's Salt) are rather rough on us. They take effect very suddenly — and, if taken too often — disrupt the body's electrolyte balance.

The most widely known laxative sugar is lactulose. It has a useful double effect, both retaining water in the colon and feeding the flora of the gut. Those little creatures can help, for example, by producing substances that act as stool softeners, or others which stimulate movement in the gut wall. But this can also cause unpleasant side effects. Overfed or misplaced bacteria can produce gases, causing cramps and flatulence.

Lactulose is formed from the milk sugar, lactose, when milk is heated to high temperatures. Pasteurisation involves heating milk briefly, so pasteurised milk contains more lactulose than raw milk. UHT (ultra-high temperature processed) milk contains more lactulose than pasteurised milk, and so on. Non-milk laxative sugars are also available, including sorbitol. Sorbitol occurs naturally in some kinds of fruit — plums, pears, and apples, for example. That is one reason for the reputation that plums have as a natural laxative, and for warnings that too much fresh apple juice can cause diarrhoea. Since human beings can barely absorb sorbitol (or lactulose) into their bloodstream, it is often used as a sweetening agent. It then appears on food labels as E420, and explains why sugar-free cough sweets, for example, always include the warning, 'Excessive consumption may have a laxative effect'. Some studies have shown that sorbitol has a similar effect to lactulose, but causes fewer side effects overall (no unpleasant wind).

Of all laxatives, the short molecular chains are most easily tolerated by the body. They have the kind of complicated names that molecular chains love — polyethylene glycol, for example, known as PEG for short. They don't disrupt the body's electrolyte balance, like salts do, and don't produce wind, like sugars do. The length of the chain is often included in the name: PG3350, for example, is a chain made up of enough atoms to give it the molecular weight 3350. It is much better than PEG150, because that compound is made of such short chains that they can be inadvertently absorbed by the gut wall. That might not necessarily be dangerous, but can certainly confuse the nerves of the gut, since polyethylene glycol is definitely not part of our natural diet.

For this reason, short chains like PEG150 are not contained in laxatives, but they are used in products such as skin cream, where they perform a very similar service — making the skin more supple. It is unlikely that they are harmful, but the matter is not finally settled. Laxatives based on PEGs contain only indigestible chains, and for that reason can be taken over longer periods without causing problems — the latest studies show there is no risk of addiction or long-term damage. Some studies indicate that these substances can even improve the gut's protective barrier.

Osmotic laxatives work not only by making the faeces moister, but also by sheer mass. The more moisture, well-fed gut bacteria, or molecular chains are contained in the gut, the more motivated it will be to move. This is the basic principle of the peristaltic reflex.

The perfect little pile by means of slippery stool

... it sounds like a children's party game: slippery stool — lots of fun, but may be quite messy. But, in fact, this is the technique known medically as faecal lubrication. Robert Chesebrough, the man who invented Vaseline, swore by a spoonful of the petroleum jelly a day. Swallowing Vaseline probably has the same effect as swallowing other fat-based faecal lubricants — an overdose of indigestible fat coats the goods in transit, making for an easier exit. Chesebrough lived to the ripe old age of 96, which is quite surprising, since eating a fat-based lubricant every day will cause the body to lose too many fat-soluble vitamins. They also get covered in the fatty lubricant, and go the same way as the faeces. This can cause vitamin deficiencies that lead to illness, especially if faecal lubricants are taken too often or in excessive amounts. Vaseline is not one of the official faecal lubricants (and really shouldn't be eaten) — but the time-honoured candidates such as liquid paraffin are hardly more unsuitable for regular use. They can be useful as short-term treatments — for example, in the presence of small but painful injuries to the anus, or haemorrhoids. In such cases, it can be good to make the faeces softer, to avoid pain or further injury during defecation. However, gel-forming fibres available from the pharmacy do just as good a job, and are less dangerous and better for the body.

The perfect little pile by means of hydragogues

... is achieved by giving the gut a big kick up the bum. These laxatives are ideal for those with very shy, lethargic gut nerves. There are various tests to find out whether that applies to you. One test involves swallowing little medical pellets that doctors then X-ray as they pass through the gut. If, after a certain time, the pellets are still spread out through the tract, having failed to gather by the back door as they should, hydragogues are the appropriate treatment.

Hydragogues latch on to a couple of the receptors that the gut

Fig.: *Hydragogues encourage the gut to keep things moving in the right direction.*

waves around inquisitively. They then send signals to the gut to stop extracting fluid from the food passing through, and to fetch more water from other parts: muscles — shake a leg! To put it bluntly, cleverly constructed hydragogues simply boss nerve cells and water transporters around. When osmotic laxatives fail to provide enough stimulus or softness, a shy gut needs a few clear commands. If taken before bedtime and left to do their thing overnight, the gut will react the next morning. If time is an issue, the express-mail service of a suppository can usually deliver the message within half an hour.

The commando squad need not rely on chemical weapons — some plants work in much the same way. These include aloe vera and senna, for example. But they do have one rather interesting side effect. Anyone who has ever wished to dye the inside of their gut black is welcome to have a go. This discoloration is not dangerous, and fades with time.

However, some scientists have reported effects of an excess of hydragogues or aloe vera that would be rather less fun, if they turn out really to be caused by those substances, as they include damage to the nerves. The reason is that nerves which are bossed around too much eventually become overwrought. When this happens, they withdraw into themselves, like snails when you tap their feelers. For this reason, patients with long-term problems should not take hydragogues more than every two or three days.

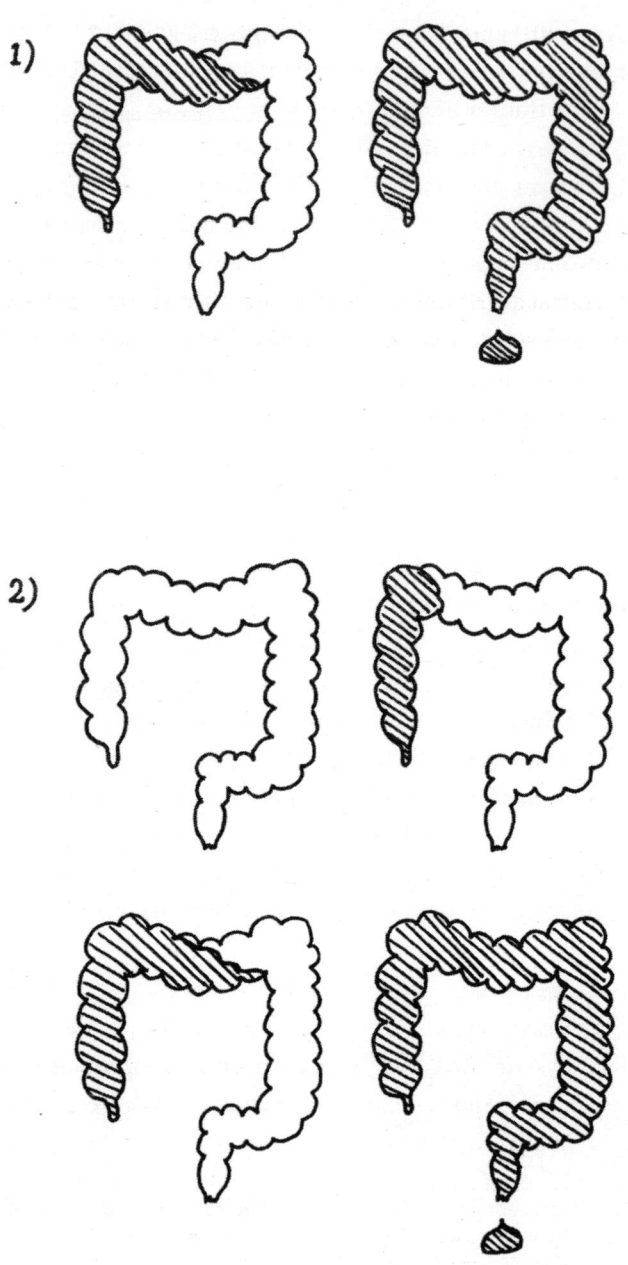

The perfect little pile by means of prokinetics

... is the very latest thing — for two reasons. These drugs can only support the gut in doing what it naturally does anyway, and cannot issue any unwanted orders. They work along the same lines as a loudspeaker. The exciting thing for many scientists is that they can help in an extremely targetted way. Some prokinetic agents affect only one single receptor; others are not absorbed into the bloodstream at all. However, scientists are still researching the way many of these substances work, and others are just now appearing on the market. So, anyone who does not absolutely need to try something new should stay on the safe side and rely on better-tested drugs for now.

The three-day rule

Many doctors prescribe laxatives without explaining the three-day rule, although it is easy to remember and is a useful aid. The large intestine has three sections: the ascending, transverse, and descending colon. When we go to the toilet, we usually empty the last section. By the next day, it has filled up again, and the game starts all over again. Taking a strong laxative may cause the entire large intestine — all three sections — to be emptied. It can then easily take three days before the large intestine is full again.

Those unfamiliar with the three-day rule will likely start to get nervous during that time. Still no stool? And before they know it, they've taken the next laxative tablet or powder. This is a vicious and unnecessary circle. After taking a laxative, the gut deserves a couple of days' respite. Monitoring for normal stool should begin on the third day. Slow transporters may need to give a helping hand to their gut after two days.

Fig.: *1. Normal situation: one-third of the large intestine is emptied, and is full again by the next day. 2. After taking a laxative: the entire large intestine is emptied, and it may take three days to fill up again.*

The Brain and the Gut

This is a sea squirt.

It may be enlightening to learn its view on the necessity of having a brain. The sea squirt, like humans, is a member of the chordate phylum. It has a bit of a brain and a kind of spinal cord. The brain blithely sends messages to the rest of the body via the spinal cord, and receives interesting information in return. In humans, for example, it might receive an image of a traffic sign from the eyes; a sea squirt's eyes might tell it when a fish swims by. A human's brain might receive information from the sensors in the skin about whether it's cold outside; a sea squirt's skin sensors can tell its brain about the temperature of the water deeper in the sea. A human might get information about whether certain foods are good to eat … and so might a sea squirt.

Equipped with all this information, a young sea squirt navigates the great oceans until it finds a rock that is secure, located in water that is just the right temperature, and surrounded by food. Having found a home, it settles down. Sea squirts are in fact sessile animals; once they take up residence, they never move again, come what may. The first thing a sea squirt does after setting up home is to eat its own brain. And why not? It's possible to live and be a sea squirt without one.

Daniel Wolpert is not only an engineer and medical doctor who has won many academic honours; he is also a scientist who believes the sea squirt's attitude to having a brain is very significant. His theory is that the only reason for having a brain is to enable movement. On first hearing, that might sound like an annoyingly banal statement. But perhaps we just consider the wrong things banal.

Movement is the most extraordinary thing ever developed by living creatures. There is no other reason for having muscles, no other reason for having nerves in those muscles, and probably no other reason for having a brain. Everything that has ever been done in the history of humankind has only been possible because we are able to move. Movement involves not just walking or throwing a ball; it is also pulling faces, uttering words, and putting plans into action. Our brain coordinates its senses, and creates experience in order to produce movement — movement of the mouth or the hands, and movement over many kilometres, or over just a few millimetres. Sometimes we can also influence the world around us by suppressing movement. But if you're a tree and can't choose whether you move or not, you don't need a brain.

The common-or-garden sea squirt no longer needs a brain after it has settled in one place. Its time of movement is over, and so its brain is surplus to requirements. Thinking *without* moving is less useful than having a mouth opening to eat plankton with. The latter influences the balance of nature at least a tiny bit.

We humans are very proud of our particularly complex brains. Thinking about constitutional law, philosophy, physics, or religion is an impressive feat, and can prompt extremely sophisticated movements. It is awe-inspiring that our brains are capable of all this. But, at some point, that awe wears off, and we hold our brains responsible for everything we experience in life — we think up experiences of wellbeing, happiness, or satisfaction inside our own heads. When we are insecure, anxious, or depressed, we worry that the computer in our heads might be broken. Philosophising and physics research *are* matters of the mind, and always will be — but there is more to our 'self' than that.

And it is from the gut that we learn this lesson: the organ that is responsible for little brown heaps, and unbidden sounds and smells of all sorts. This is the organ that is currently forcing researchers to rethink its role; in fact, scientists are cautiously beginning to question the view that the brain is the sole and absolute ruler of the body. The gut not only possesses an unimaginable number of nerves, but those nerves are also unimaginably different from those of the rest of the body. The gut commands an entire fleet of signalling substances, nerve-insulation materials, and ways of connecting. There is only one other organ in the body that can compete with the gut for diversity — the brain. The gut's network of nerves is called the 'gut brain' because it is just as large and chemically complex as the grey matter in our heads. Were the gut solely responsible for transporting food and producing the occasional burp, such a sophisticated nervous system would be an odd waste of energy. No body would create such a neural network just to enable us to break wind. There must be more to it than that.

We humans have known since time immemorial something that science is only now discovering: our gut feeling is responsible in no small measure for how we feel. We are 'scared shitless' or we can be 'pooing our pants' with fear. We can't get our 'arse into gear' if we don't manage to complete a job. We 'swallow' our disappointment and need time to 'digest' a defeat. A nasty comment leaves a 'bad taste in the mouth'. When we fall in love, we get 'butterflies in our stomach'. Our 'self' is created in our head and our gut — no longer just in language, but increasingly also in the lab.

How the gut influences the brain

When scientists study feelings, they start out by looking for something to measure. They draw up scales for suicidal tendencies, test hormone levels to measure love, or trial tablets to treat anxiety. To outsiders, this often appears less than romantic. In Frankfurt, there was even a study which involved scanning the brains of volunteers while a research assistant tickled their genitals with a toothbrush. Such experiments tell scientists which areas of the brain receive signals from which parts of the body. This helps them draw a map of the brain.

So they now know, for example, that signals from the genitals are sent to the upper, central part of the brain, just below the crown. Fear is found in the middle of the brain — right between the ears, so to speak. Word formation is located just above the temple. Morality is located behind the forehead, and so on. In order to improve our understanding of the relationship between the gut and the brain, we must trace their communication pathways. How do signals get from belly to brain, and what effect do they have when they get there?

Signals from the gut can reach different parts of the brain, but they can't reach everywhere. For example, they never end up in the visual cortex at the back of the brain. If they did, we would see visual effects or images of what is going on in our gut. Regions they can end up in, however, include the insula, the limbic system, the prefrontal cortex, the amygdala, the hippocampus, and the anterior cingulate cortex. Any neuroscientists reading this will be up in arms when I roughly define the responsibilities of these brain regions as, respectively, self-awareness, emotion, morality, fear, memory, and motivation. This does not mean that our guts control our moral thinking, but it allows for the possibility that the gut might have a certain influence on it. Scientists need to conduct more laboratory experiments to look more closely at this possibility.

The *forced-swimming test,* carried out on mice, is one of the most revealing experiments performed in the name of research into motivation and depression. A mouse is placed in a small container of water that is too deep for it to reach the bottom of with its feet, forcing it to swim around and around, trying in vain to get to dry land. The question is, how long will it keep swimming in pursuit of its aim? This boils down to one of the basic questions of our existence: how intensely are we prepared to strive for something that we believe exists? This might be something concrete, like dry land beneath our feet, or high-school graduation. Or it might be something abstract, like satisfaction or happiness.

Mice with depressive tendencies do not swim for long. They simply freeze, apathetically awaiting their fate. It seems that inhibitory signals are transmitted more efficiently in their brains than motivational or driving impulses. Such mice also show a stronger reaction to stress. New antidepressants can normally be tested on these mice. If they swim for longer after receiving the medication, it is an indication that the substance under scrutiny might be effective.

Researchers in the team led by the Irish scientist John Cryan took this one step further. They fed half their mice with *Lactobacillus rhamnosus JB-1*, a strain of bacteria known to be

---

Fig.: *Areas of the brain activated by vision, fear, word formation, moral thought and genital stimulation.*

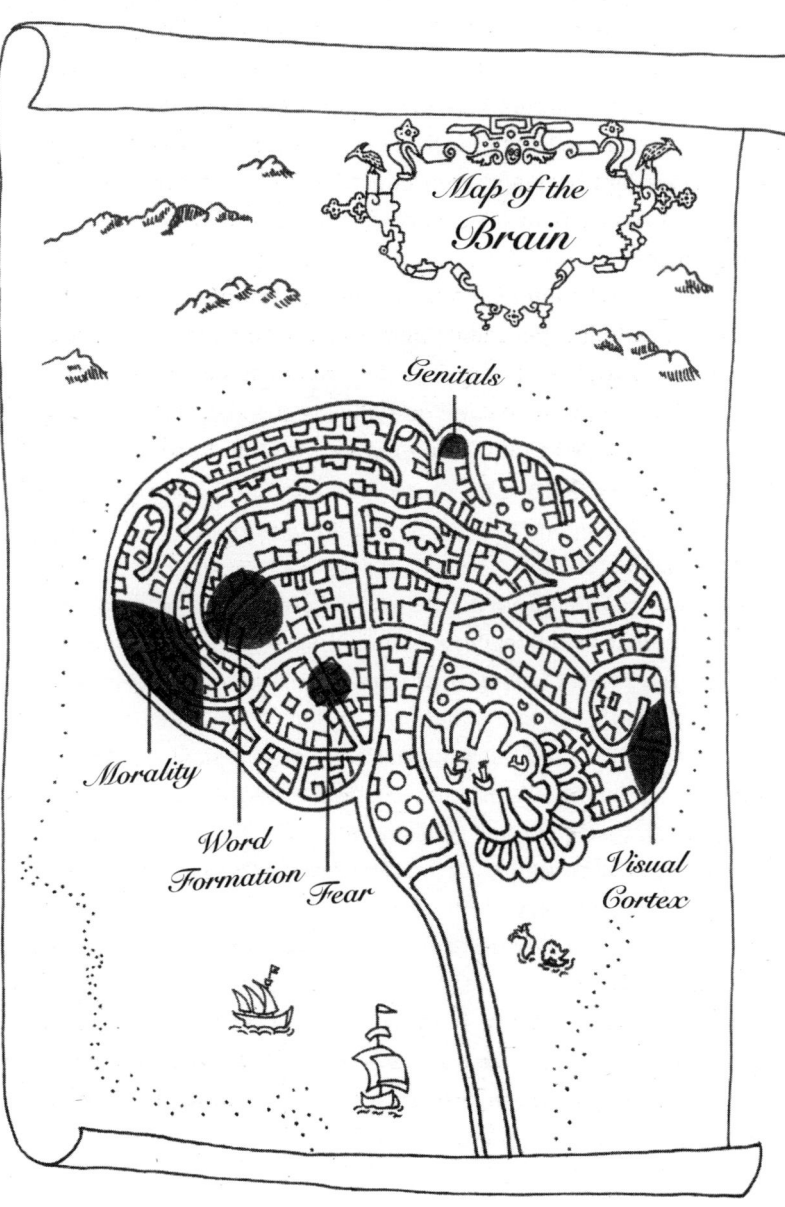

good for the gut. Back in 2011, the idea of altering the behaviour of mice by changing the contents of their gut was very new. And, indeed, the mice with the 'pimped' gut flora not only kept swimming for longer and with more motivation, but their blood was also found to contain fewer stress hormones. Furthermore, these mice performed better in memory and learning tests than their un-pimped peers. When scientists severed their vagus nerve, however, no difference was recorded between the two groups of mice.

This nerve is the fastest and most important route from the gut to the brain. It runs through the diaphragm, between the lungs and the heart, up along the oesophagus, through the neck to the brain. Experiments on humans have shown that they can be made to feel comfortable or anxious by stimulating the vagus nerve at different frequencies. In 2010, The European Union approved a medical treatment that uses stimulation of the vagus nerve to help patients suffering from depressive disorders. So, this nerve works something like a telephone connection to the switchboard at a company's headquarters, transferring messages from staff out in the field.

The brain needs this information to form a picture of how the body is doing. This is because the brain is more heavily insulated and protected than any other organ in the body. It nestles in a bony skull, surrounded by a thick membrane, and every drop of blood is filtered before it is allowed to flow through the regions of the brain. The gut, by contrast, is right in the thick of it. It knows all the molecules in the last meal we ate, inquisitively intercepts hormones as they swim around in the blood, enquires of immune cells what kind of day they're having, and listens attentively to the hum of the bacteria in the gut. It is able to tell the brain things about us it would never otherwise have had an inkling of.

The gut has not only a remarkable system of nerves to gather all this information, but also a huge surface area. That makes it the body's largest sensory organ. Our eyes, ears, nose, or skin pale by comparison. The information they gather is received by the conscious mind, and used to formulate a response to our environment. They can be seen as life's parking sensors. The gut, by contrast, is a huge matrix, sensing our inner life and working on the subconscious mind.

Cooperation between the gut and the brain begins very early in life. Together, they are responsible for a large proportion of our emotional world when we are babies. We love the pleasant feeling of a full stomach, get terribly upset when we are hungry, or grizzle and moan with wind. Familiar people feed, change, and burp us. It's palpably clear that our infant 'self' consists of the gut and the brain. As we get older, we increasingly experience the world through our senses. We no longer scream blue murder when we don't like the food at a restaurant. But the connection between gut and brain does not disappear overnight; it simply becomes more refined. A gut that does not feel good might subtly affect our mood, and a healthy, well-nourished gut can discreetly improve our sense of wellbeing.

The first study of the effect of intestinal care on healthy human brains was published in 2013 — two years after the study on mice. The researchers assumed there would be no visible effect on humans. The results they came up with were surprising — not only for them, but for the entire research community. After four weeks of taking a cocktail of certain bacteria, some of the areas of the subjects' brains were unmistakably altered, especially the areas responsible for processing emotions and pain.

Of irritated bowels, stress, and depression

Not every unchewed pea can intervene in the brain's activity. A healthy gut does not transmit minor, unimportant digestive signals to the brain via the vagus nerve. Rather, it processes them with its own brain — that's why it has one, after all. If it thinks something is important, however, it may consider calling in the brain.

By the same token, the brain does not transfer every piece of information to the conscious mind. If the vagus nerve wants to deliver information to the extremely important locations in the brain, it must get them past the doorman, so to speak. The brain's bouncer is the thalamus. When our eyes report to the thalamus for the twentieth time that the same curtains are still hanging at the living-room window, it refuses entry to that information — it is not important for the conscious mind. A report of *new* living-room curtains is something it would let in, for example. That is not true of everybody's thalamus, but most people's.

An unchewed pea will not make it across the threshold from the gut to the brain. The story is different for other stimuli, however. For example, a report of an unusually large intake of alcohol will make it from the belly to the head, where it informs the vomit-control centre; information about trapped gas will reach

the pain centre; and the presence of pathogenic substances will be reported to the officer in charge of nausea. These stimuli make it through because the gut's threshold and the brain's doorman consider them important. But it is not only bad news that makes it across the border. Some signals can cause us to fall asleep on the couch, contented and full after a big Christmas dinner. We are conscious that some of these signals originate from the abdomen; others are processed in the subconscious areas of the brain, and so cannot be located so clearly.

When a gut is irritated, its connection to the brain can make life extremely unpleasant. This shows up on brain scans. In one experiment, the activity in the brains of volunteers was imaged while a small balloon was inflated inside their intestine. Healthy subjects showed normal brain activity, with no notable emotional components. When patients with irritable bowels were subjected to the same procedure, however, there were clear indications of activity in the emotional centre of the brain normally associated with unpleasant feelings. So the stimulus was able to cross both barriers in those subjects. The patients felt uneasy, although they had not endured anything untoward.

Irritable-bowel syndrome is often characterised by an unpleasant bloated feeling or gurgling in the abdomen, and a susceptibility to diarrhoea or constipation. Sufferers also have an above-average incidence of anxiety or depressive disorders. Experiments like the one with the balloon show that feeling unwell and experiencing negative emotions can arise via the gut-brain axis — when the gut's threshold is lowered, or the brain insists on having information it would not normally receive.

Such a state of affairs may be caused by tiny but persistent (micro-) inflammations, bad gut flora, or undetected food intolerances. Despite a wealth of recent research, some doctors still dismiss patients with irritable-bowel syndrome as hypochondriacs or malingerers, because their tests show no visible damage to the gut.

Other diseases affecting the bowel are different. During an acute phase of their condition, patients with a chronic inflammatory bowel disease like *Crohn's disease* or *ulcerative colitis* may have real sores in their bowel wall. With these conditions, the trouble is not that even tiny stimuli are transferred from the gut to the brain — their threshold is still high enough to prevent that. The problems are caused by the diseased mucous membrane of the gut. Like patients with irritable-bowel syndrome, sufferers of these conditions also show increased rates of depression and anxiety.

There are currently very few — but very good — research teams studying how to make the threshold between the gut and the brain less porous. This is important not only for patients with intestinal problems, but for all of us. Stress is thought to be among the most important stimuli discussed by the brain and the gut. When the brain senses a major problem (such as time pressure or anger), it naturally wants to solve it. To do so, it needs energy, which it borrows mainly from the gut. The gut is informed of the emergency situation via the sympathetic nerve fibres, and is instructed to obey the brain in this exceptional period. It is kind enough to save energy on digestion, producing less mucous and reducing the blood supply.

However, this system is not designed for long-term use. If the brain constantly thinks it is in an emergency situation, it begins to take undue advantage of the gut's compliance. When that happens, the gut is forced to send unpleasant signals to the brain to say it is no longer willing to be exploited. This negative stimulus can cause fatigue, loss of appetite, general malaise, or diarrhoea. As with emotional vomiting in response to upsetting situations, the gut reacts by ridding itself of food to save energy so it is available to the brain. The difference is that real stress situations can continue for much longer than minor upsets do. If the gut has to continue to forego energy in favour of the brain, its health will eventually suffer. A reduced blood supply and a thinner protective layer of mucous weaken the gut walls. The immune cells that dwell in the gut wall begin to secrete large amounts of signal substances that make the gut brain increasingly sensitive and lower the first threshold. Periods of stress mean the brain borrows energy; and, as any housekeeper knows, good budgeting is always better than running up too many debts.

One theory proposed by research bacteriologists is that stress is unhygienic. The altered circumstances that stress creates in the gut allow different bacteria to survive there than in periods of low stress. We could say that stress changes the weather in the gut. Tough guys who have no problem with turbulence will reproduce successfully — and, at the end of the day, they are not likely to spread good cheer in the gut. If this theory is true, that would make us not just the victims of our own gut bacteria, but also the gardeners of our inner world. It would also mean that our gut is capable of making us feel the negative effects, long after the period of stress is over.

Feelings from down below, especially those that leave a nasty aftertaste, will cause the brain to think twice next time about whether it really wants to hold a speech in front of the entire office, or whether we really should eat that super-hot chilli. So the process of making decisions based on 'gut feeling' may involve the gut recalling how it felt in similar situations in the past. If positive lessons could also be reinforced in the same manner, then the way to a lover's heart really would be through their stomach — and straight to the gut.

The interesting theory that our gut is involved not only in our feelings and in making 'gut decisions', but may also influence our behaviour, is the subject of various research projects. A team led by Stephen Collins designed an ingenious experiment using two different strains of mice with very well-researched behavioural characteristics. Members of the strain called BALB/c are more timid and docile than those belonging to the NIH-SWISS strain,

which exhibit more exploratory behaviour and gregariousness. The researchers gave the mice a cocktail of antibiotics that only affect the gut, wiping out their entire gut flora. They then fed the animals with gut bacteria that were typical of the other strain. Behaviour tests showed they had swapped roles — the BALB/c mice became more gregarious, and the NIH-SWISS mice were more timid. This shows that the gut can influence behaviour — at least in mice. The result cannot yet be applied to humans. Scientists know far too little about the various bacteria involved, about the gut brain in general, and about the gut-brain axis.

Until scientists have filled in those gaps in their knowledge, we can make use of the facts we already know to improve gut health. It starts with the little things like mealtimes, for example, which should be enjoyed without pressure, at a leisurely pace. The dinner table should be a stress-free zone, with no place for scolding or pronouncements like 'You will remain at the table until you've finished the food on your plate!', and without constant TV channel hopping. This is important for adults, but it is vital for small children, whose gut brain develops in parallel with their head brain. The earlier in life that mealtime calm is introduced, the better. Stress of any kind activates nerves that inhibit the digestive process, which means we not only extract less energy from our food, but we also take longer to digest it, putting the gut under unnecessary extra strain.

We can play around with this knowledge and test it experimentally. Tablets or medicated chewing gum to prevent travel sickness work by numbing the nerves of the gut. When the nausea abates, feelings of anxiety often disappear, too. If unaccountable grumpiness or anxiety can originate in the gut (even without nausea), is it possible that these drugs could be used to banish them? By temporarily numbing a troubled tummy, so to speak? Alcohol reaches the nerves of the gut before it reaches those in the brain — so how much of the relaxing effect of that 'just one glass of wine' in the evening actually comes from a sedated gut brain? What about the array of bacteria in the wide range of yoghurt on our supermarket shelves? Is *Lactobacillus reuteri* better for me than *Bifidobacterium animalis*? A team of Chinese researchers managed to show in the laboratory that *Lactobacillus reuteri* is able to inhibit pain sensors in the gut.

Lactobacillus plantarum and *Bifidobacterium infantis* could already be recommended as a pain treatment for patients with irritable-bowel syndrome. Many patients with a low pain threshold in the gut currently take substances designed to treat diarrhoea, constipation, or cramps. That might help with the symptoms, but does not address the cause of the problem. If, after eliminating possible food intolerances and re-stocking the gut flora, there is still no improvement, we must take the problem by the scruff of its neck — or, in this case, by the nerve-cell threshold. So far, very few treatments have been scientifically proven to be effective. Those which have include cognitive behavioural therapy and hypnotherapy.

Really good psychotherapy is like physiotherapy for the nerves. It eases tensions, and teaches us how to move in a more healthy way — at the neural level. Because the nerves of the brain are more complicated creatures than muscles are, a trainer needs to have more creative exercises up his sleeve. Hypnotherapists often use thought journeys and guided-imagery techniques, which aim to reduce the intensity of pain signals, and alter the way the brain processes certain stimuli. Just like muscles, certain nerves can become stronger with increased use. The therapy does not involve hypnotism as seen in TV shows. That would, in fact, be self-defeating, since this kind of therapy relies on the patient being in control at all times. Patients must make sure the hypnotherapist they choose is recognised by a reputable institution.

Hypnotherapy has been shown to be effective in treating patients with irritable-bowel syndrome, reducing their reliance on medication — in some cases, to zero. This is particularly true of children with this condition. For them, hypnotherapy has been shown to produce a 90 per cent reduction in pain, compared to a 40 per cent reduction produced by drugs. Some clinics (including the Saarbrucken Clinic in Germany) offer specific hypnotherapy for abdominal complaints.

Patients with intestinal disease who also suffer from extreme anxiety or depressive disorders are often recommended antidepressants by their doctor. However, they are rarely told why. And there is a simple reason for that: no doctor or scientists knows. It was not until they noticed the mood-enhancing effects of these drugs that scientists began to explore the mechanisms behind this phenomenon. They still have not come up with a clear answer. For decades, it was thought to be due to an enhancing effect on the 'happiness hormone', *serotonin*. More recently, depression researchers have also begun investigating another possibility — that such drugs may increase the plasticity of the nerves.

Neuroplasticity is the nerves' ability to change. It is nerve plasticity that makes puberty such a confusing time for an adolescent brain — so much is still being moulded into shape. The possibilities are endless, and nerves are constantly firing off messages in all directions in a pubescent brain. This process is not complete until we reach the age of about 25. After that, nerves react according to well-rehearsed patterns. Patterns that have proved useful in the past are retained; others are rejected as failures. This explains the disappearance not only of the inexplicable fits of laughter and temper tantrums of our teenage years, but also the posters on our bedroom walls. After this age, we find it more difficult to deal with sudden change, but the payback is a more stable, calmer disposition. This can also result in negative thought patterns taking root, such as 'I am worthless' or 'Everything I do goes wrong.' The nervous messages from a worried gut can also become embedded in a person's mind. If it is the case that antidepressants increase neuroplasticity, they may work by loosening up such negative thought patterns. This is most beneficial when accompanied by effective psychotherapy to help patients resist slipping back into old habits.

The side effects of commercially available antidepressants, such as Prozac, also provide us with important clues about the 'happiness hormone', *serotonin*. A quarter of the patients will report typical side effects such as nausea, an initial phase of diarrhoea, and constipation when the drug is taken over a long period of time. This is explained by the fact that our gut brain possesses the same neural receptors as the brain in our head. So, antidepressants automatically 'treat' both brains. The American researcher Dr Michael Gershon takes this line of thought one stage further. He is interested in the possibility of developing effective antidepressants that only influence the gut, and do not have an effect on the brain.

That is not as outlandish as it might first seem — after all, 95 per cent of the *serotonin* we produce is manufactured in the cells of our gut, where it has an enormous effect on enabling the nerves to stimulate muscle movement, and acts as an important signalling molecule. If its effects on the gut can be changed, the messages sent from there to the brain would also be changed enormously. This would be particularly useful in treating the sudden onset of severe depression in people whose lives are otherwise fine. Perhaps their gut needs a session on the therapist's couch, and their head is not to blame at all.

Anyone who suffers from anxiety or depression should remember that an unhappy gut can be the cause of an unhappy mind. Sometimes, the gut has a perfect right to be unhappy, if it is dealing with an undetected food intolerance, for example. We should not always blame depression on the brain or on our life circumstances — there is much more to us than that.

Where the 'self' originates

Grumpiness, happiness, insecurity, wellbeing, and worry do not originate in isolation in the mind. We are human beings, with arms and legs, genitals, a heart, lungs, and a gut. Science's concentration on the brain has long blinded us to the fact that our 'self' is made up of more than just our grey matter. Recent gut research has contributed significantly to a new, cautious questioning of the philosophical proposition, 'I think, therefore I am'.

One of the most fascinating parts of the brain that can receive information from the gut is the insula, or insular cortex. This part of the brain is being studied by one of the most brilliant brains working in research today: Bud Craig. With superhuman patience, he has spent the last twenty years staining nerve fibres and tracking their paths through the brain. Eventually, he emerged from the lab and gave a one-hour talk revealing his theory that human self-awareness originates in the insular cortex.

The first part of his hypothesis goes like this: the insula receives information about feelings from the entire body. Each piece of information is like a pixel — and the insula organises these pixels to form an overall image. This image is important because it represents a map of our feelings. So, when we are sitting on a chair, we feel the cheeks of our bottom pressed against the seat, and we might also feel cold, or hungry, for example. Taken together, this gives the overall picture of a cold, hungry person sitting on a hard chair. We might not find this image particularly great, but it is also not awful — it's just okay.

Hypothesis, part two: Daniel Wolpert tells us that the purpose of the brain is to create movement — irrespective of whether you are a sea squirt searching for a comfy rock beneath the sea, or a human being striving for the best life possible. The aim of movement is to bring about an effect. The brain can use the insula's map to plan meaningful movement. If I am sitting around feeling cold and hungry, other areas of the brain will be motivated to do something to change my situation. I could start shivering, or get up and head for the fridge in search of food. One of the main purposes of movement is to shift us constantly towards a healthy equilibrium — from cold to warm, from sad to happy, or from tired to alert, for example.

Hypothesis, part three: the brain is an organ of the body. So, if the insula creates an image of the body, that image must also include the on-board computer in our head. It includes some interesting areas, such as those responsible for social empathy, morality, and logic. The social areas of the brain might induce negative feelings when we argue with our partner, or make our logic regions despair when we try to solve a difficult puzzle. In order for the insula to create a reasonable image of our self, it probably also takes in perceptions of our environment and experiences from the past. So, when we are cold, we don't just feel the low temperature; we are able to contextualise the feeling, and think such thoughts as: *This is weird. I'm cold, but I'm in a well-heated room indoors. Maybe I'm coming down with something?* Or, alternatively, *Okay, maybe I shouldn't be parading around naked in the conservatory in winter.* In this way, humans are able to react to the stimulus of feeling cold in a much more complex way than other animals.

The more information we connect, the cleverer the movements we can make. In this respect, there is probably also a hierarchy among our organs. Information that is particularly important for the maintenance of a healthy equilibrium has more sway in the insula. The brain and the gut are well qualified to take a central role — if not *the* central role.

So the insula creates a picture of our entire feeling body. We can then use our complex brain to embellish this image. Bud Craig believes the picture is 'refreshed' approximately every forty seconds. Through time, those images merge into a kind of movie — the film of the self, of our life.

A great deal of what makes up this movie is certainly contributed by the brain — but not everything. It may be time to expand René Descartes' proposition along these lines: 'I feel, then I think, therefore I am.'

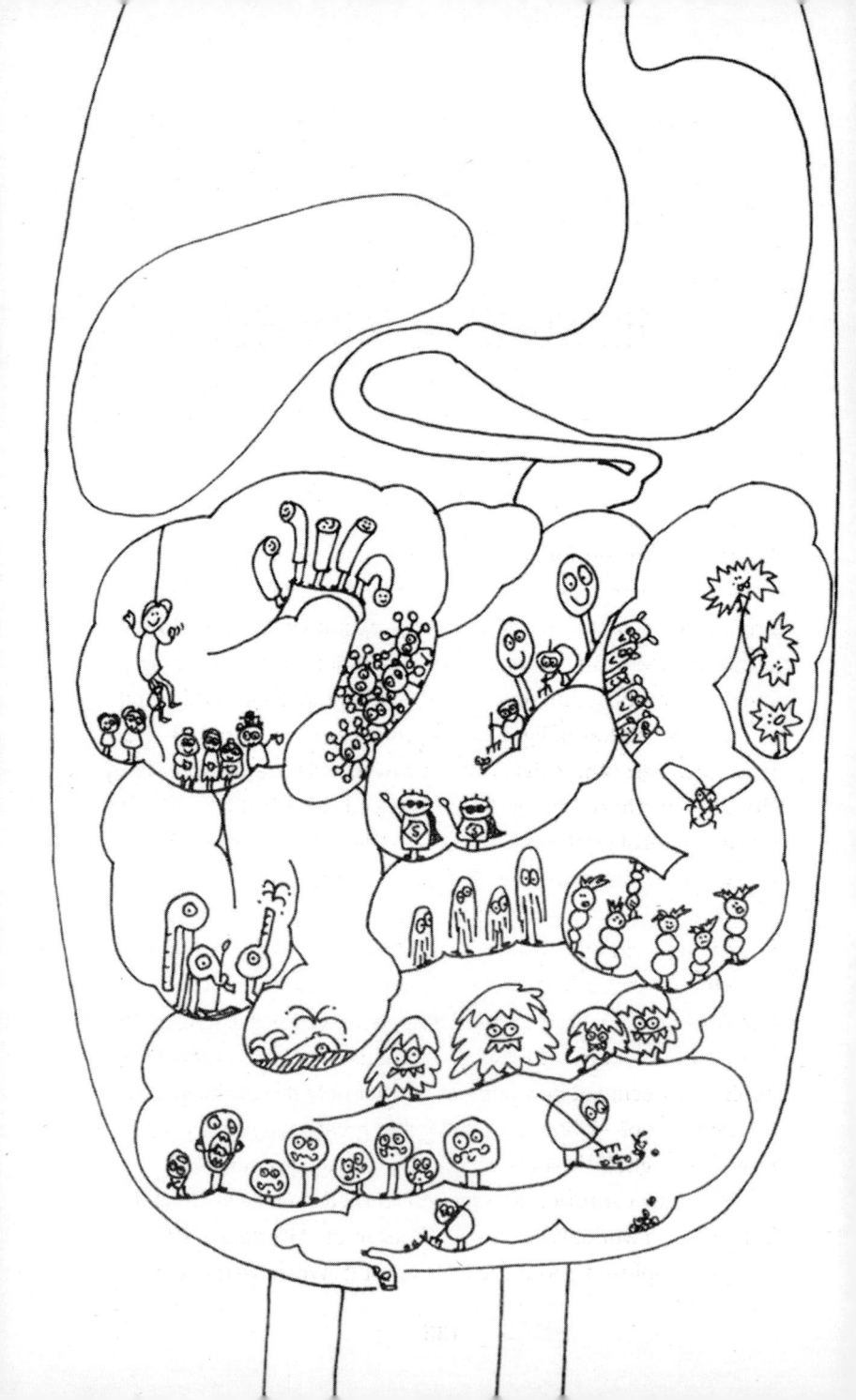

3
THE WORLD OF MICROBES

Looking down on Earth from space, it is impossible to make out any human beings. The Earth itself is easily identified — a bright, round spot among the other bright spots in the darkness of space. Closing in a little, it becomes apparent that we humans inhabit all sorts of different places on the planet. Our cities shine at night as patches of light. Some groups are concentrated in big urban centres; others live scattered over wide areas. We inhabit the cool northern climes of Europe and America, but we also occupy tropical rainforests and the margins of arid deserts. We are everywhere, even though we are invisible from space.

Looking closer at human beings, it becomes clear that each of us is a world of our own. Our forehead is a breezy meadow, our elbows are arid wastelands, our eyes are salty lakes, and our gut is the most amazing giant forest ever, populated by the weirdest of creatures. Just as we humans occupy the planet, our bodies are occupied by a population that only reveals itself under the microscope — bacteria. Viewed at great magnification, they resemble bright little spots against a dark background.

For many centuries, humans concerned themselves with the human-sized world. We measured it, examined its flora and fauna, and philosophised about life as lived in it. We constructed huge machines and flew to the Moon. Explorers keen to discover new continents today must turn to the microscopic world within us. Our gut is perhaps the most fascinating continent of that world. It provides the habitat for more species and families of creatures than any other landscape. Exploration of this region is only now beginning in earnest. A new sense of excitement is brewing among scientists, not unlike that associated with the decoding of the human genome and all the promise that holds for the future. Of course, this excitement about gut research could still fizzle and come to nothing.

Work on an atlas of human bacteria did not begin until 2007. This project involves taking cotton-swab samples from all over the bodies of lots and lots of people. Samples are collected from three different areas of their mouths, under their armpits, and on their foreheads. Stool samples are analysed, and genital smears are evaluated. Places once thought to be bacteria-free turn out to be populated after all — the lungs, for example. When it comes to drawing up a bacteria atlas, the gut is the supreme challenge. Of our entire microbiome — that is, all the micro-organisms that teem on the inside and outside of our bodies — 99 per cent are found in the gut. Not because there are so few elsewhere, but because there are simply so inconceivably many in the gut.

I Am an Ecosystem

We are familiar with bacteria — the little creatures consisting of a single cell. Some live in boiling hot springs in Iceland; others luxuriate in the moisture on a dog's wet nose. Some require oxygen to produce energy, and 'breathe' almost like us. Others die when exposed to the fresh air. They draw their energy not from oxygen, but from metal atoms or acids — and that can result in some rather interesting smells. Almost every human body smell is, in fact, produced by bacteria. From the pleasant scent of a loved-one's skin to the dog-breath of next-door's frisky hound — it is all the product of the microscopic world in us, on us, and around us.

We like to watch athletic surfers as they ride the waves, but we are unaware of the spectacular surfing scenario that takes place in our nasal flora every time we sneeze. We pant and sweat when we exercise, but no one notices how delighted the bacteria who live in our running shoes are over the sudden, summery change in climate. When we take a sneaky bite of cake, we never hear the roar of excited bacteria in our gut, as they gleefully shout 'HERE COMES CA-A-A-KE!' It would take an entire international news agency to report on all the events constantly unfolding in just one person's microbiome. While we lounge about feeling bored, any number of exciting things are happening inside us.

People are slowly beginning to realise that the vast majority of bacteria are harmless, or even helpful. Some facts are now known to science. Our gut's microbiome weighs around two kilos and contains about 100 trillion bacteria. One gram of faeces contains more bacteria than there are people on the Earth. We also know that this community of microbes crack open indigestible foodstuffs for us, supply the gut with energy, manufacture vitamins, break down toxins and medications, and train our immune system. Different bacteria manufacture different substances: acids, gases, fats — bacteria are tiny factories. We know that gut bacteria are responsible for blood groups, and that harmful bacteria cause diarrhoea.

What we don't know is what all this means for each individual. We notice pretty quickly when we have ingested diarrhoea-causing bacteria. But what do we notice of the work carried out by the many millions, billions, trillions of tiny creatures inside us every day? Does the precise nature of the creatures that colonise us make a difference? Skewed proportions of the different bacteria in our gut have been detected in those suffering from obesity, malnutrition, nervous diseases, depression, and chronic digestive problems. In other words, when something is wrong with our microbiome, something goes wrong with us.

One person might have stronger nerves than another because she has a better stock of vitamin B-producing bacteria. Another person might be able to deal easily with bit of bread mould eaten by mistake, or yet another might have a tendency to gain weight because the 'chubby bacteria' in his gut feed him a bit too willingly. Science is just beginning to understand that each of us is an entire ecosystem. Microbiome research is still young, complete with wobbly milk teeth and short pants.

When scientists still knew very little about bacteria, they classified them as plants. This explains terms like gut 'flora', which is not scientifically accurate, but appropriately descriptive. A bit like plants, different bacteria have different characteristics concerning their habitat, nutrition, or level of toxicity. The scientifically correct terms are microbiota [little life] and microbiome to refer to our collection of microbes and their genes.

In general, it is accurate to say that their numbers are smaller

Fig.: *Bacteria population density in the different regions of the gut.*

in the upper sections of the digestive tract, while a very, very large number of bacteria reside in the lower parts, such as the large intestine and the rectum. Some prefer the small intestine; others live exclusively in the colon. There are great fans of the appendix, well-behaved homebodies that stick to the mucous membrane, and rather cheekier chaps that nestle up close to the cells of our gut.

It is not always easy to get to know our gut microbes personally. They don't like to be removed from their own world. When scientists try to grow them in the lab to observe them, they simply refuse to cooperate. Skin bacteria merrily gobble up the lab food and grow into little microbe mountains — gut bacteria don't. More than half the bacteria that grow in our digestive tract are just too well adapted to living there to be able to survive outside the gut. Our gut is their world. It keeps them warm, moist, protected from oxygen, and supplied with pre-tasted food.

Only ten years ago, many scientists would probably have maintained that there is a stable stock of gut bacteria that is more or less common to every human being. For example, when they spread faeces on a culture medium, they always found *E. coli* bacteria. It was as simple as that. Today, we have machines that can scan a gram of faeces, molecule by molecule. This reveals the genetic remains of billions of bacteria. We now know that *E. coli* makes up less than 1 per cent of the population in the gut. Our gastrointestinal tract is home to more than a thousand different species of bacteria — plus minority populations of viruses and yeasts, as well as fungi and various other single-celled organisms.

You might think that our immune system would pounce on this multitude of settlers. Defending the body from foreign invasions is high on the immune system's to-do list. Sometimes it even wages war on tiny pollen grains that accidently get sucked into our nostrils. Hay fever sufferers know the signs: a streaming nose and itchy eyes. So how do the bacteria side-step the immune system and stage a bacterial Woodstock inside our bodies?

The Immune System and Our Bacteria

We face potential death every day. We could get cancer, get eaten away by bacteria, or become infected with a deadly virus. And several times a day, our lives are saved. Strangely mutated cells are destroyed, fungal spores are eliminated, bacteria are peppered with holes, and viruses are sliced in two. This agreeable service is provided by our immune system, with its many little cells. Its workforce includes experts at spotting foreign bodies, contract killers, 'hatters' (more on this later), and mediators. They all work hand in hand, and are a pretty top team.

The vast majority of our immune system (about 80 per cent) is located in the gut. And with good reason. This is where the main stage at the bacterial Woodstock is situated, and any immune system worth its salt must be there, or be square. The bacteria are confined to a fenced-off area — the mucous membrane of the gut — preventing them from getting too dangerously close to the cells of the gut wall. The immune system is able to play with the cells without ever posing a danger to the body. This allows our defender cells to get acquainted with many previously unknown species.

If, sometime later and elsewhere in the body, an immune cell encounters a now-familiar bacterium, it can react to it much more quickly. The immune system has to be extremely careful in the gut, suppressing its defensive instincts and allowing the many bacteria there to live in peace. But, at the same time, it must also recognise dangerous elements in the crowd, and weed them out. If we decided to say 'Hi' to each of our gut bacteria individually, we might just manage it in around 3 million years. Our immune system not only says 'Hi'; it also says, 'You're okay', or 'I'd prefer to see you dead.'

Also, strange as it may sound, the immune system must be able to distinguish between bacterial cells and the body's own human cells. That is easier said than done. Some bacteria have structures on their surface that bear a close resemblance to those on the surface of our own cells. This is the reason why, for example, scarlet fever should be treated immediately with antibiotics. If it is not treated quickly, the immune system may begin to mistake the cells of the joints or other organs for the bacteria that cause scarlet fever, and attack them. It might suddenly think that our knee is a nasty sore-throat germ hiding out in our leg. It happens rarely — but it does happen.

Scientists have observed a similar effect in patients with juvenile diabetes. Also called diabetes mellitus type 1, this condition results from the autoimmune destruction of the cells that produce insulin. One possible cause is a breakdown of communication with the bacteria of the gut. It may be that they are failing to train the immune system properly, or the immune system is somehow getting the message wrong.

The body has a very rigorous set of measures designed to guard against such communication breakdowns and cases of mistaken identity. Before an immune cell is allowed to enter the bloodstream, it has to complete the toughest boot camp of any cells. Among other things, it must cover a vast distance while being confronted constantly with various structures of the body. If our little immune cell encounters something it cannot clearly identify as belonging to the body or coming from outside, it stops and prods at it a little. That is a fatal error: this cell will never graduate to the bloodstream.

In this way, immune cells with a tendency to attack the body's own tissue are weeded out before they leave boot camp. In their training centre in the gut, they learn to be tolerant of foreign bodies; or, rather, they learn to be better prepared for an encounter with them. This system works rather well, and usually without untoward incidents.

There is one lesson that is particularly tricky to learn: what to do about foreign bodies that aren't actually bacteria, but remind the immune system of them? Red-blood cells, for example, have very bacteria-like proteins on their surface. Our immune system would attack our own blood if it had not learned in boot camp that the blood is a no-go area. If our blood cells have the blood-group marker A on their surface, we encounter no problem receiving transfusions of blood from donors with the same blood group. The reasons for needing a transfusion can be diverse, from a motorbike accident to heavy blood-loss during childbirth.

However, we cannot receive blood from donors whose blood cells have a different blood-group marker on their surface. It would immediately remind our immune system of bacteria, and since the immune system knows that bacteria have no business being in the bloodstream, it would consider the donated blood cells an enemy, and cause the cells to form clumps. If it weren't for this combat readiness — learned through training by our gut bacteria — there would be no blood groups, and any donor could give blood to any recipient. For newborn babies, who do not yet have many bacteria in their guts, this is indeed the case. They can theoretically receive transfusions of blood from any group, without any incompatibility effects. (As a precaution, hospitals give babies blood from the mother's blood group, since antibodies from the mother can find their way into the baby's bloodstream.) As soon as babies begin to develop a rudimentary immune system and gut flora, they can only tolerate blood from their own group.

Blood-group development is just one of many immunological phenomena caused by bacteria. There are probably many more waiting to be discovered. Much of what bacteria do tends to be 'fine tuning'. Each kind of bacterium has its own way of affecting the immune system. Some species have been observed to make our immune system more tolerant, for example, by causing more peace-loving, mediatory immune cells to be produced, or by affecting our cells in a similar way to cortisone and other anti-inflammatory drugs. That results in a milder, less belligerent immune system. This is probably a clever move on the part of these tiny creatures, since it increases their chances of being tolerated in the gut.

The fact that the small intestines of young vertebrates (including humans) have been found to contain bacteria that provoke the immune system leaves room for speculation. Could it be the case that these provocateurs help keep the bacteria population in the small intestine down? That would make the small intestine an area of low bacteria tolerance, giving it some peace and quiet for a while. The provocateurs themselves do not hang out in the mucous membrane like good little bacteria, but dock firmly onto the villi of the small intestine. A similar preference is shown by pathogens, such as harmful versions of *E. coli*. When they want to colonise the small intestine, but find their favourite places occupied by such provocateurs, they have no choice but to leave.

This effect is known as colonisation resistance. The majority of the microbes in our gut protect us simply by occupying spaces that would otherwise be free for harmful bacteria to colonise. Incidentally, the provocateurs of the small intestine belong to that group of characters that refuse to be cultivated outside the gut. How can we be sure that they are not harming us? Well, we can't. It is possible that they do harm some people by overstimulating the immune system. Many questions remain to be answered.

Some of those questions might be answered with the help of a group of germ-free mice in a New York laboratory. They are the cleanest creatures in the world: the product of sterile caesarean births, antiseptic cages, and steam-sterilised food. Disinfected

---

Fig.: *If antibodies match with foreign blood cells, the cells will form clumps. Blood group B has antibodies against blood group A.*

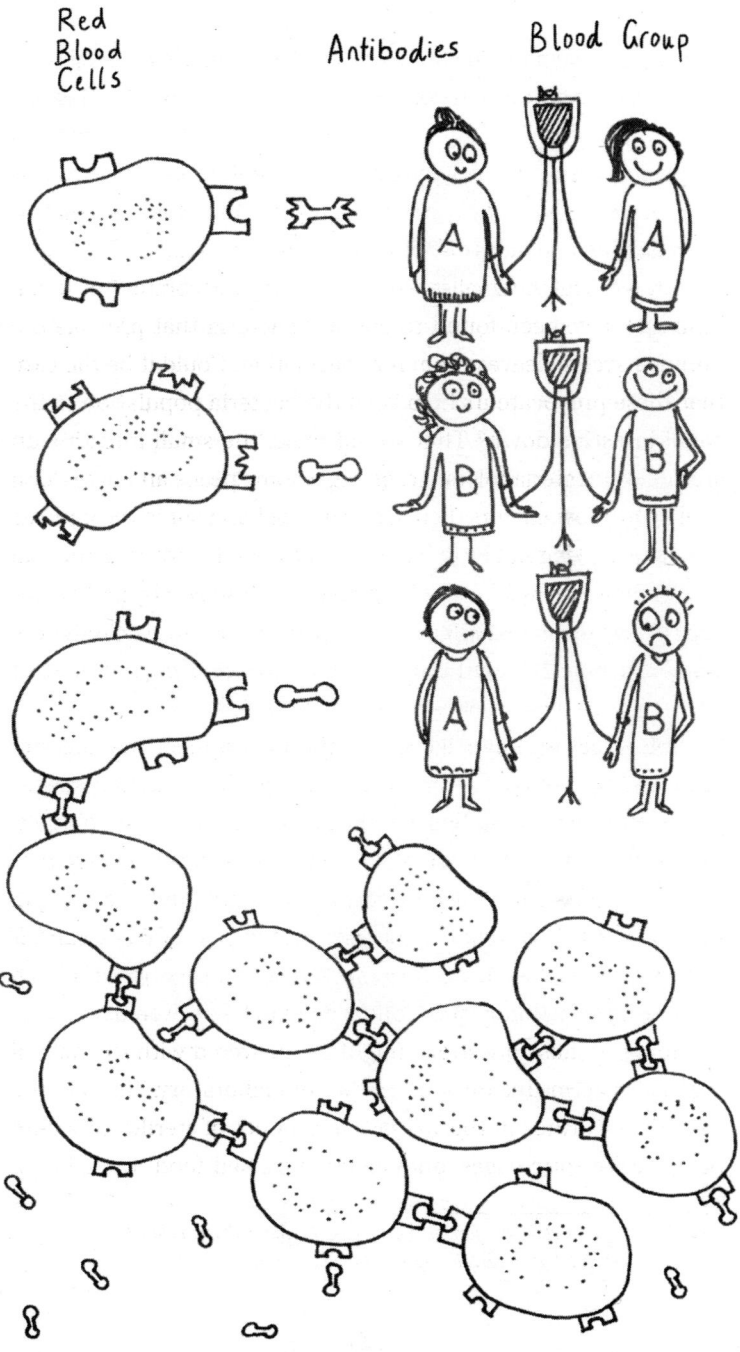

animals like these can never exist in nature. Anyone working with these mice must take the utmost care, since even unfiltered air teems with flying germs. The mice allow researchers to watch what happens to an immune system that has nothing to do. What goes on in a gut with no microbes? How does an untrained immune system react to pathogens? What differences are so obvious they can be seen by the naked eye?

Anyone who has ever had anything to do with such animals will tell you that germ-free mice are weird. They are often hyperactive, and exhibit an un-mouse-like lack of caution. They eat more than their germ-colonised peers, and take longer to digest their food. They have hugely enlarged appendices, shrivelled digestive tracts with few villi and blood vessels, and a reduced number of immune cells. They are easy prey for even relatively harmless pathogens.

Feeding them with cocktails of bacteria taken from other mice produces astonishing results. If they are given bacteria from mice with type 2 diabetes, they soon begin to develop problems metabolising sugar. If bacteria from obese humans are fed to germ-free mice, they are more likely to gain weight than if they receive bacteria from people in the normal weight range. Scientists can also administer a single species of bacteria to observe its effect on the mice. Some bacteria, acting alone, are able to reverse the effects of the sterile environment — cranking up the immune system, shrinking the mice's swollen appendices down to normal size, and normalising their eating behaviour. Other lone bacteria have no effect whatsoever. Yet others take effect only in cooperation with colleagues from other bacteria families.

Studies using these mice have advanced our knowledge quite considerably. We now have good reason to speculate that, just as the macroscopic world we live in influences us, we are also influenced by the microscopic world that lives in us. This is all the more interesting when we realise that each person's inner world is unique to him or her.

The Development of the Gut Flora

As unborn babies, we live in an environment that is normally completely germ-free — the womb. For nine months, we have no contact with the outside world, except through our mother. Our food is pre-digested; our oxygen is pre-breathed. Our mother's lungs and gut filter everything before it reaches us. We eat and breathe through her blood, which is kept free of germs by her immune system. We are sheathed in an amniotic sac and encased in a muscly uterus, which is corked with a thick plug like a big earthenware jug. All this means that not a single parasite, virus, bacterium, fungus — and certainly no other person — can touch us.

This situation is unusual. Never again in our lives will we be so protected and so isolated. If we were designed to remain germ-free once we leave the womb, we would be very different creatures. But that is not the case, so all living things of any size have at least one other living thing that helps them in some way, and is allowed to live on or in them in return. This explains why our cells are constructed in such a way that bacteria can easily dock with structures on their surface, and it explains why certain bacteria have co-evolved with us over many millennia.

As soon as there is any breach in the protective amniotic sac, colonisation begins. While 100 per cent of the cells that make us up when we start life are human cells, we are soon colonised by so many micro-organisms that only 10 per cent of our cells are human, with microbes accounting for the remaining 90 per cent. We cannot see this, because our human cells are so much larger than those of our new lodgers. Before we look into our mother's

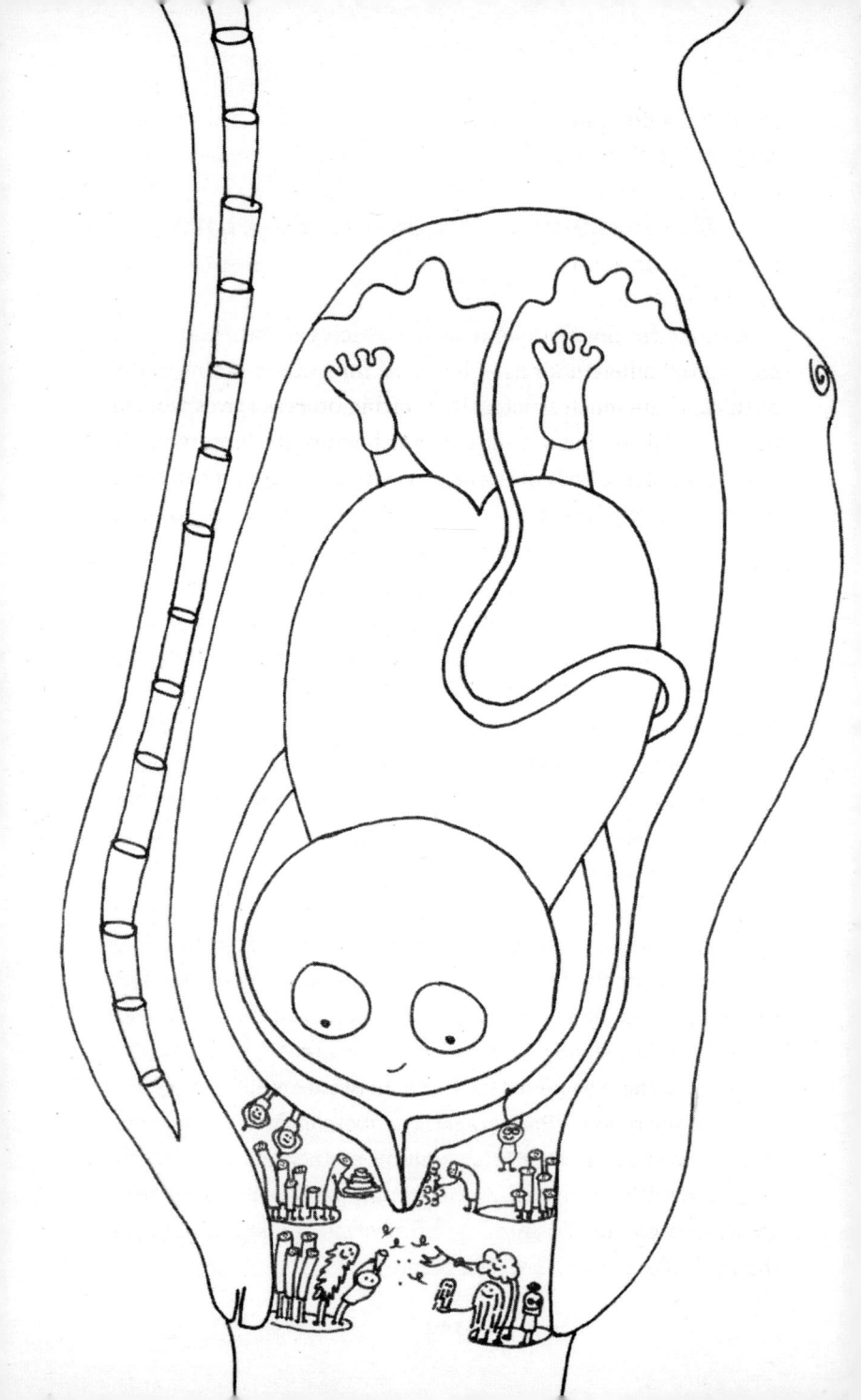

eyes for the first time, the creatures that live in her body cavities have already looked into ours. The first ones we meet are her protective vaginal flora — an army that defends a very important territory. One way it does this is by producing acids that drive away other bacteria and keep the way cleaner and cleaner, the closer they are to the womb.

Unlike the flora in our nostrils, which can be made up of around 900 different kinds of bacteria, the criteria for life in the birth canal are much stricter. This sorting process leaves women with a useful coating of bacteria, which wraps itself protectively around the sterile body of the baby as it emerges. About half these bacteria are from one genus: lactobacillus. Their favourite pastime is producing lactic acid. This means the only residents that can set up home in the birth canal are those that pass the acid test.

With a routine birth, all we have to do as babies is decide which way to face as we come out. There are two attractive possibilities: towards the back or towards the front. During birth we are exposed to all sorts of skin contact before we are wrapped up in something soft by another person, usually wearing latex gloves.

By now, the founding fathers of our first microbial colonies are already in us and on us: mainly our mother's vaginal and gut flora, mixed with a few skin-dwelling germs, and possibly a few others from the hospital's repertoire. This is a very good mixture to start with. The acid army protects us from harmful invaders while other bacteria are already starting to train the immune system, and the first, indigestible components of our mother's milk are broken down for us by helpful microbes.

Some of these bacteria take less than twenty minutes to spawn the next generation. What takes us twenty years or more, happens in a fraction of the time — a fraction as tiny as the colonists themselves. While our first gut bacterium watches its great-great-great-great grandchild swim by, we have spent just two hours in the arms of our proud new parents.

Despite this rapid population growth, it will be about three years before the gut flora develop to the right level and then stabilise. Prior to that, our tummies are the scene of dramatic power struggles and great bacterial battles. Some folks who find their way into our mouth spread rapidly throughout the gut, only to disappear again just as quickly. Others will remain with us for the rest of our lives. The composition of our gut colony depends partly on our own actions: we might lick our mother's skin, gnaw on a chair leg, and give the car window or the neighbour's dog the occasional sloppy kiss. Anything that finds its way into our mouths in the process could soon be building its empire in the world of our gut. Whether it will continue to prevail will remain to be seen. And whether its intentions are good or bad will also remain to be seen. So we gather our own fate with our mouths, so to speak. Stool samples are able to show what comes out the other end. It's a game with many unknowns.

We receive some help creating this collection — mainly from our mothers. No matter how many sloppy kisses we give the car window, if we are allowed to kiss and cuddle with our mothers regularly, we will be protected by her microbes. Breastfeeding also promotes particular members of our gut flora — breast-milk-loving *Bifidobacteria*, for example. Colonising the gut so early, these bacteria are instrumental in the development of later bodily functions, such as those of the immune system or the metabolic system. Children with insufficient *Bifidobacteria* in their gut in their first year have an increased risk of obesity in later life, compared with infants with large populations.

There are many, many kinds of bacteria — some beneficial, others less so. Breastfeeding can help shift the balance towards the beneficial, and reduce the risk of a later gluten intolerance, for example. A baby's first population of gut bacteria prepares the way for the later 'adult' population by removing oxygen and electrons from the intestine. As soon as the environment is free of oxygen, the more typical bacteria of the gut can start to settle there.

Breast milk is so beneficial that a more-or-less well-nourished mother need not do any more than suckle her baby to ensure it is receiving a healthy diet. When it comes to the nutrients it contains, breast milk provides everything that dietary scientists believe children need in order to thrive — like the best dietary supplement ever. It contains everything, knows everything, and can do everything necessary for a child's wellbeing. And, as if that weren't enough, it has the added advantage of passing on a bit of mum's immune system to her offspring. Breast milk contains antibodies that are able to protect against any dangerous bacteria a child might make the acquaintance of (by licking the family pet, for example).

Weaning is the first revolution experienced by a baby's gut flora. Suddenly, the entire composition of junior's food supply is different. Clever old Mother Nature has equipped the typical bacteria that first colonise the infant gut with the genes needed to break down simple carbohydrates such as those in rice. If you serve junior with complex plant-based foods like garden peas, for example, his baby flora will not be able to deal with them alone. He is now going to need new kinds of digestive bacteria. African children have bacteria that can manufacture all kinds of tools needed to break down even the most fibrous of plant-based foods. The microbes in the guts of European children prefer to avoid such hard work — and they may do so with a clear conscience, since their diet consists primarily of puréed baby food and small amounts of meat.

Bacteria do not always manufacture the tools they need; sometimes they also borrow them. In Japan, the (gut) population has entered into a trade relationship with marine bacteria. They borrowed a gene from their sea-living colleagues that helps break down the kind of seaweed used in Japanese cuisine to make sushi, for example. This shows that the composition of our gut population can depend to a large extent on the tools we need to break down certain foodstuffs.

Useful gut bacteria can be passed on through the generations. Anyone of European heritage who has experienced constipation after a blow-out session at the all-you-can-eat sushi bar will appreciate the advantage of inheriting Japanese seaweed-processing bacteria from someone in the family. However, it is not so easy to infuse yourself or your kids with a few sushi-digesting assistants. Bacteria have to like living in the place where they work.

If we say that a micro-organism is particularly suited to our gut, we mean that it appreciates the architecture of our gut cells, copes well with the climate, and likes the food on the menu. All three of these factors vary from person to person. Our genes help design our bodies, but they are not the chief architects of our microbial home. Identical twins share the same genes, but do not have the same bacterial mix. They do not even have noticeably more similarities than other pairs of siblings. Our lifestyle, random acquaintances, illness, or hobbies all influence the shape of the populations inside our bodies.

On the way to a relatively mature gut flora in our third year, we stick all sorts of stuff in our mouths — some of which will be useful and suited to us. We acquire more and more micro-organisms, building up our population diversity from a couple of hundred species of bacteria to many hundreds of different gut-dwellers. That would be a pretty impressive inventory for any zoo, yet we acquire this variety without even thinking about it.

It is now generally accepted that the first populations to colonise our gut lay the main foundations for the future of our entire body. Studies have shown the importance of those first few weeks of postnatal bacteria-collecting for the development of the immune system. Just three weeks after birth, the metabolic products of our gut flora may predict our increased risks of allergies, asthma, or neurodermatitis in later life. How do we manage to pick up bacteria that are more harmful than beneficial to us, so early in life?

For a long time, researchers focussed on one factor in particular: more than a third of all children in Western, industrialised countries are brought into the world by means of a convenient caesarean section. No squeezing through a narrow birth canal, no unpleasant side effects such as perineal tearing, no delivering the afterbirth — it sounds like a fine thing. However, children born by caesarean section have a higher risk of picking up typical hospital germs and are more likely to develop allergies during the course of their lives. Initial studies show that their gut flora are considerably altered. And researchers were quick to identify this as a possible cause of all manner of ills.

Luckily, more precise research methods today mean we now know children born by caesarean section receive the beneficial bacteria they need from their mother in other ways. Within just a few days, they accumulate a similar number of maternal 'starter germs' to babies who enter the world via vaginal birth.

This means scientists must look elsewhere for new sources of damaging gut bacteria. Unbeneficial early combinations of bacteria in the gut can also have other causes — poor nutrition, unnecessary use of antibiotics, excessive cleanliness, or exposure to bad bacteria can also feature among the factors. In spite of all this, there is no reason for anyone to feel inadequate — we humans are such large creatures, there is no way we can exercise control over every aspect of the microscopic world.

The Adult Gut Population

In terms of our microbiota, we reach 'adulthood' around the age of three. For a gut, being an adult means knowing how you work and what you like. When that stage has been reached, some gut microbes find themselves on a great expedition with us through our entire lives. We dictate the itinerary — by eating what we eat, exposing ourselves to stress or avoiding it, by going through puberty, by getting ill and by growing old.

Those people who post pictures of their dinner on Facebook, only to be disappointed by the lack of 'likes' from friends, are simply trying to appeal to the wrong audience. If there were such a thing as 'Facebug' (Facebook for microbes!), a picture of your dinner would provoke an excited response from millions of users, and shudders of disgust from millions more. The menu changes daily: useful milk digesters contained in a cheese sandwich; armies of salmonella bacteria hiding in a delicious dish of tiramisu. Sometimes we alter our gut flora, and sometimes they alter us. We are our flora's weather and their seasons. Our flora can take care of us, or they can poison us.

We are only now beginning to learn the impact that the belly-based bacteria community can have on an adult human. In this respect, scientists know more about bees than human beings. For bees, having more diverse gut bacteria has been a successful evolutionary strategy. They were only able to evolve from their carnivorous wasp ancestors because they picked up new kinds of gut microbes that were able to extract energy from plant pollen. This allowed bees to become vegetarians. Beneficial bacteria provide bees with an insurance policy in times of food scarcity:

they have no trouble digesting unfamiliar nectar from far-flung fields. More specialised digesters are not so well equipped. Times of crisis highlight the advantage of hosting a good microbial army: for example, bees with well-equipped gut flora can deal with parasite attacks better than those without. Gut bacteria are an incredibly important factor in this evolutionary survival-strategy.

Unfortunately, we cannot simply transfer these results to humans. Humans are not bees; they are vertebrates, and they use Facebook. So researchers have to go back to square one. Scientists investigating our gut bacteria have to learn to understand an almost completely unknown world and its interaction with the world outside. First, they need to know who is living inside our gut.

So let's take a closer look. Who exactly are these characters?

Biologists love to put things in order — from the contents of their own desks to the entire contents of the world. They begin by sorting everything into two large drawers: one for living things, the other for non-living things. They then go on to divide everything in the first drawer into three categories: *eukaryote*s, archaea, and bacteria. Representatives of all three groups can be found in the gut. I am not promising too much when I say that each of the three groups has its own kind of charm.

Eukaryotes are made up of the largest and most complex cells. They can be multicellular, and grow to a pretty impressive size. A whale is a *eukaryote*. Humans are *eukaryote*s. Ants are too, incidentally, although they are much smaller than we are. Modern biologists divide *eukaryote*s into six subgroups: crawly amoeboid microbes, microbes with 'pseudopodia' (foot-like protrusions that aren't real feet), plant-like organisms, single-celled organisms with little mouth-like feeding grooves, algae, and *opisthokonts*.

For those unfamiliar with the term *opisthokont* (it comes from the Greek words for 'rear' and 'pole'), it describes the group that includes all animals — humans as well — and also fungi. So the next time you meet an ant in the street, you can give it a friendly wave as a fellow *opisthokont*. The most common *eukaryote*s found in the gut are yeasts, which are also *opisthokonts*. We are familiar with yeast as a rising agent for bread, but there are many other kinds.

Archaea are kind of in-betweeners — not really *eukaryote*s, but not really bacteria, either. Their cells are small and complex. If this description seems a little vague, it may help if I say that the *Archaea* are pretty rad characters. They love the extreme things in life. Some are *hyperthermophiles*, which feel right at home in temperatures of more than 100°C and are often to be found hanging out near volcanoes. The *acidophiles* among the *Archaea* like to paddle around in highly concentrated acid. *Barophiles* (also called *piezophiles*) thrive under pressure, and have specially adapted cell walls to allow them to live on the deep sea floor. *Halophiles* are at home in extremely salty water (they love the Dead Sea). The rare characters among the *Archaea* that can be cultured in the lab are the *cryophiles*, which love the cold. They like laboratory freezers that keep them at a cosy -80°C. There is one species of *Archaea* often found in our gut that thrives on the waste products of other gut bacteria, and can glow.

Returning to the main topic: bacteria make up more than 90 per cent of the population of our gut. Biologists divide bacteria into more than twenty phyla, or lineages. Members of different phyla are sometimes about as similar as human beings are to *excavates* (the single-cell microbes with the feeding grooves) — that is, not very. Most of the inhabitants of our gut belong to one of five phyla: mainly *Bacteroidetes* and *Firmicutes*, with a smattering of *Actinobacteria*, *Proteobacteria*, and *Verrucomicrobia*. These phyla are further divided into increasingly specific categories, until we eventually reach the level of the bacteria family. Members of a given family are relatively similar to each other. They eat the same food, keep similar company, and have similar abilities. Individual family members have impressive names like *Bacteroides uniformis*,

Lactobacillus acidophilus, or *Helicobacter pylori*. The bacteria kingdom is huge.

Whenever scientists search humans for a particular bacterium, they constantly come across new, previously unknown species. Or they discover known species in unexpected places. In 2011, a group of researchers in the US decided to examine the flora of volunteers' belly buttons, just for fun. One subject's navel was found to contain bacteria that were previously known to live only in the seas off the coast of Japan — despite the fact that the volunteer had never been to Asia. Globalisation is not just your local corner shop turning into a MacDonald's — it affects even the contents of our navels. Every day, billions and billions of foreign micro-organisms fly round the world, without paying a single cent for their tickets.

Everyone has his or her personal collection of bacteria. It could even be described as a unique bacterial fingerprint. If you were to take swabs from a dog and analyse the genes of its bacteria, the dog's owner could be easily identified with reasonable certainty. The same is true of computer keyboards. Anything that we come into regular contact with carries our microbial signature. Everyone has some outlandish items in their bacterial collection that almost no one else will share.

The microbial landscape in our gut is just as unique and individual. So how are doctors supposed to know what is beneficial and what is harmful? Uniqueness like this presents researchers with a problem. If they are trying to ascertain what influence our gut bacteria have on our health, it is no use just finding out that Mr Smith is carrying a strange Asian species and several other weird microbes in his gut. Scientists need to identify patterns to deduce facts from them.

Fig.: *Rough overview of the three most important phyla of bacteria and their subgroups. Lactobacilli are Firmicutes, for example.*

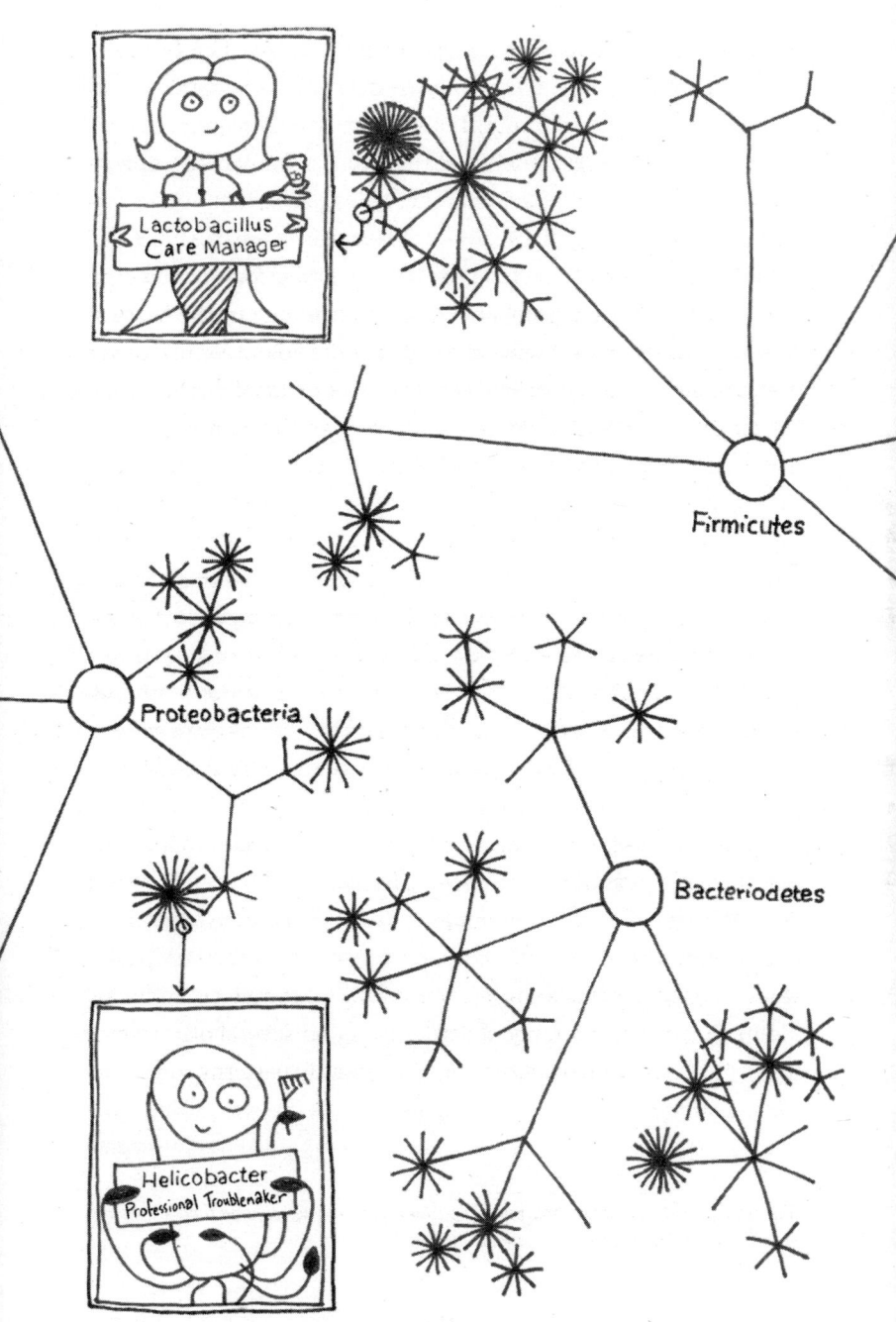

So, since scientists are faced with more than a thousand different families of bacteria, they must decide whether they just need to identify rough lineages, or whether they should look at every paid-up member of the *Bacteroides* bacteria family individually. *E. coli* and its evil twin, EHEC, are members of the same family, for example. The differences between them are infinitesimally small — but they are very tangible: *E. coli* is a harmless gut-dweller, while EHEC causes severe internal bleeding and diarrhoea. It always makes sense to examine families or lineages when you want to know what damage an individual bacterium can wreak.

The genes of our bacteria

Genes are possibilities. Genes are information. Genes can be dominant, forcing features on us, or they can just offer their abilities for us to use or not. But, most of all, genes are plans. They are incapable of doing anything unless they are read and implemented. Implementation of some of these plans is obligatory — for one, they decide whether we are a human being or a bacterium. Others may be left on the back burner for years (liver spots, for example), and yet others might be carried for a lifetime without ever being expressed — as in the case of genes for large breasts. Some people might consider that a pity; others, a blessing.

Taken together, our gut bacteria have 150 times more genes than a human being. This massive collection of genes is called a biome. If we could pick 150 different living things, parts of whose genetic blueprints we would like to possess, what would we choose? Some people might opt for the strength of a lion, the wings of a bird, the hearing of a bat, or the practical mobile home of a snail.

There are many reasons why it would be more practical to opt for bacteria genes instead. They can be taken in easily via the mouth, can unfurl their abilities in the gut, and can even adapt to our lifestyle. Nobody needs a snail's mobile home all the time, and no one needs breast-milk digestion helpers forever. The latter disappear gradually after weaning. It is not yet possible to examine all the genes of our gut bacteria at once. However, it is possible to search specifically for individual genes, if you know what you are looking for. We know that babies contain more active genes for digesting breast milk than adults do. The guts of obese people are often found to contain more bacterial genes involved in breaking down carbohydrates. Older people have fewer bacterial genes for dealing with stress. In Tokyo, they can help digest seaweed; in Pforzheim, they probably can't. Our gut bacteria paint a rough portrait of who we are: young, chubby, or Asian, for example.

The genes of our gut bacteria also inform us about our body's abilities. The pain-relief drug paracetamol can be more toxic for some people than others: some gut bacteria produce a substance that influences the liver's ability to detoxify the drug. Whether you can pop a pill to cure your headache without a second thought is decided partially in your gut.

Similar caution should be exercised with general dietary tips. Soya's ability to protect against prostate cancer, cardiovascular disease, or bone disorders, for example, has now been proven. More than 50 per cent of Asians benefit from this effect. But among people of European heritage, this beneficial effect is only found among 25 to 30 per cent of the population. This cannot simply be explained by human genetic differences. The difference is due to certain bacteria. They are found more commonly in the guts of Asians, and are able to coax the health-promoting essence out of tofu and other soya products.

It is great for science when it identifies the individual bacterial genes that are responsible for this beneficial effect. In such cases, science can be said to have come up with an answer to the question of how gut bacteria influence our health. But we want more than that. We want to understand the big picture. If you look at all the bacterial genes so far discovered, the small individual groups of genes responsible for breaking down painkillers or soya products fade into the background. The common features they share dominate the picture: every microbe contains many genes involved in breaking down carbohydrates or proteins, and in producing vitamins.

Science has the same problem when investigating the micro biome that the Google generation regularly faces. We ask a question, and six million sources send us simultaneous answers. We don't respond by telling them to form an orderly queue. We have to sort them shrewdly into categories, weed out the irrelevant ones, and recognise important patterns. One important step in this direction was the discovery of the three human *enterotypes*, in 2011.

Researchers in Heidelberg, Germany, were using cutting-edge technology to investigate the human gut biome. They expected to see the usual picture: a chaotic mixture of many different bacteria, including a host of unknown species. What they actually discovered came as a surprise. Despite the great diversity, there was order. One of three families was always dominant in the realm of bacteria. Suddenly, the whole mess of more than a thousand families looked much more organised.

The three gut types

A person's enterotype depends on the family of bacteria that dominates the microbe population of their gut. The choice is between families that bask in the glory of the names *Bacteroides*, *Prevotella*, and *Ruminococcus*. Researchers identified these enterotypes distributed among Asians, Americans, and Europeans, irrespective of age or gender. In the future, enterotyping may help doctors predict a whole range of characteristics, such as the body's response to soya, nerve resilience, or susceptibility to certain diseases.

Practitioners of traditional Chinese medicine who were visiting the institute in Heidelberg at the time of this discovery recognised an opportunity to combine their ancient knowledge with modern medicine. Classical Chinese medical theory has always divided people into three groups, according to how they react to certain medicinal plants, such as ginger. The families of bacteria in our bodies also have different characteristics. They break down food in different ways, produce different substances, and detoxify certain harmful substances but not others. Furthermore, they may also influence the gut flora by either encouraging or attacking bacteria from the other two groups.

Bacteroides

Bacteroides are the best-known family of gut bacteria, and often form the dominant population. They are experts in breaking down carbohydrates, and possess a huge collection of genetic blueprints allowing them to manufacture any enzyme they need to accomplish that task. Whether we eat a steak or a large salad, or chew on a raffia doormat in a drunken stupor, *Bacteroides* know straightaway which enzymes they need. Come what may, they are equipped to extract energy from it.

Their ability to extract the maximum energy from everything and pass it on to us has led to the suspicion that they may be responsible for an increased tendency to gain weight. Indeed, *Bacteroides* do seem to like meat and saturated fatty acids. They are more common in the guts of people who eat plenty of sausages and the like. But does having them in our gut make us fat, or does being fat lead to having them in our gut? This question remains to be answered. *Bacteroides* carriers are also likely to have a weakness for their colleagues, *Parabacteroides*. These bacteria are also particularly deft at passing on as many calories to us as possible.

This enterotype is also noticeable, among other things, for

its ability to produce particularly large amounts of biotin. Other terms for biotin include vitamin B7 and vitamin H. It was given the name vitamin H in the 1930s because of its ability to *heal* a certain skin condition caused by consuming too much raw egg white. 'H is for healing' might not be the most creative mnemonic, but can be useful nonetheless.

Vitamin H neutralises avidin, a toxin found in raw eggs. It causes the skin disease in question by binding strongly with vitamin H, leaving the body deficient in that substance. So, eating raw eggs causes vitamin H deficiency, which in turn can lead to skin disease.

I do not know who was eating enough raw eggs in the 1930s to lead to the discovery of this connection. I do, however, know who might possibly end up eating so much avidin in the future that they could have problems with vitamin H — pigs that accidently roam into a field of genetically modified maize. Genetic engineers have created transgenic maize with a gene that produces avidin, to make it less susceptible to insect damage during storage. When pests, or stray pigs, consume the maize, they are poisoned. When the maize is cooked, it is no longer toxic, just like a good hardboiled egg.

Another indication that our gut microbes produce vitamin H is the fact that some people excrete more of it than they take in. Since no human cell is capable of producing this substance, the only possible explanation for this is that our bacteria are functioning as hidden vitamin H factories. Vitamin H is not only necessary for 'healthy-looking skin, shiny hair, and strong nails', as you might read on the packages of supplements you can buy in your local pharmacy. Biotin is also involved in some of the body's vital metabolic processes. We need it to synthesise carbohydrates and fats for our body, and to break down proteins.

Skin, hair, and nail problems are not the only effects of biotin deficiency. It can also cause depression, lethargy, susceptibility to infections, neurological disorders, and increased cholesterol levels in the blood. However, let me issue a serious WARNING here: the list of symptoms caused by any vitamin deficiency is formidable. Most people reading them feel the symptoms apply to them in some way. But it is important to remember that you can catch a cold, or feel a bit lethargic, without jumping to the conclusion that you have a biotin deficiency. And cholesterol levels are more likely to be raised by eating a big plate of bacon for breakfast than by eating the avidin in an undercooked egg.

However, some people in higher-risk groups may well consider the possibility of a biotin deficiency. That includes anyone who takes antibiotics for an extended period; heavy drinkers; anyone who has had part of their small intestine removed; anyone reliant on dialysis; and people on certain kinds of medication. These people require more biotin than they can get from a normal diet. One 'healthy' higher-risk group is pregnant women: developing babies use up biotin like ageing refrigerators gobble up electricity.

So far, no scientific studies have been carried out to investigate how much biotin our gut bacteria provide us with. We know that they produce some, and that antibacterial medications such as antibiotics can cause biotin deficiency. Investigating whether members of the *Prevotella* enterotype are more likely to suffer from a biotin deficiency than someone of the *Bacteroides* enterotype would be a pretty exciting research project. But since the existence of the three different enterotypes was only discovered in 2011, there are many more pressing questions we need to answer.

It is not only their good 'output' that makes *Bacteroides* so successful; they also work hand in hand with others. Some species in the gut make a living by clearing away the waste left by *Bacteroides*. This is a win-win situation: *Bacteroides* work better in tidy surroundings, and the waste-disposal organisms have a secure source of income. On a different level, we find the composters, which not only utilise waste products for their own ends, but also use them to make products that *Bacteroides* can use in turn. For some metabolic pathways, *Bacteroides* themselves take on the role of composter: if they need a carbon atom to modify a molecule, they simply reach up and grab it out of the atmosphere in the gut. They always find what they are looking for, since carbon is a waste product of our metabolism.

Prevotella

In many ways, the *Prevotella* family is the opposite of *Bacteroides*. Studies have shown that they are more common in the guts of vegetarians, but they also appear in moderate meat-eaters and convinced carnivores. Our diet is not the only factor that influences the colonisation of our gut. But more about that presently.

Prevotella also have a group of bacterial colleagues they prefer working with: *Desulfovibrionales*. They often have a long flagellum — a whip-like tail used to propel them along — and so, like *Prevotella*, they are adept at trawling through our mucous membranes looking for useful proteins. They then either eat those proteins, or use them to build who-knows-what. *Prevotella* produce sulphur compounds when they work. Their smell is familiar — we know it from boiled eggs. If it weren't for *Desulfovibrionales* whipping around with their propeller-tails, snapping up the sulphur, *Prevotella* would soon find themselves drowning in a sulphur swamp of their own making. Incidentally, this gas is not dangerous to human health. Our nose wrinkles at it as a precautionary measure, as it can be toxic at concentrations thousands of times greater ...

Another substance that contains sulphur and has an interesting smell is the vitamin associated with this enterotype: thiamine. Also known as vitamin B1, this is one of the most widely recognised and important vitamins. Our brains require it not only to keep the nerves well nourished, but also to coat them in an electrically insulating layer of fat. This explains why a thiamine deficiency may be the cause of muscle tremors and forgetfulness.

A very serious lack of vitamin B1 causes a disease called beriberi. It was described in Asia as early as AD 500. Beriberi means 'I cannot, I cannot' in the Sinhalese language of Sri Lanka, and refers to the fact that sufferers have difficulty walking, due to nerve damage and muscle atrophy. It is now known that polishing rice removes the vitamin B1 it contains, and that a diet made up predominantly of this kind of rice leads to an onset of symptoms within a few weeks.

While not resulting in serious neurological or memory disorders, a less-severe vitamin B1 deficiency can cause irritability, frequent headaches, and lack of concentration. More advanced cases may cause a susceptibility to oedema and heart problems. But once again, beware: these symptoms can have many causes. They are only a reason for concern when they are unusually frequent or severe. They are rarely caused exclusively by a vitamin deficiency.

Studying the symptoms of vitamin deficiencies provides a useful insight into the part played by vitamins in certain processes. Anyone whose diet does not consist exclusively of polished white rice or alcohol is usually well supplied with the right vitamins. The fact that our gut bacteria help supply us with essential vitamins means they are far more than just a load of flagellating sulphur-poopers — and that is what makes them so fascinating.

Ruminococcus

Opinions are divided on this family — scientific opinions, at any rate. Some scientists who decided to investigate the existence of enterotypes found *Prevotella* and *Bacteroides*, but no *Ruminococcus* group. Others swear that this third group exists, and yet others insist that there is even a fourth group, or fifth group, or more. Such a state of affairs can really ruin the coffee break at a medical congress.

Let's agree for argument's sake that there is at least a possibility that this group exists. Its proposed favourite food is the cell walls of plants. Possible colleagues: *Akkermansia* bacteria, which break down the mucins in mucous, and absorb sugar pretty quickly. *Ruminococcus* produces a substance called haem, which the body needs for many things, including the production of blood.

One character who probably had problems producing haem was Count Dracula. A genetic defect has been identified in his home country, Romania, which results in symptoms that include a lack of tolerance to garlic, sensitivity to sunlight, and the production of red urine. This urine discolouration is caused by a defect in blood production, which means that sufferers excrete the unfinished precursors of blood production. Nowadays, those affected by the condition, called porphyria, are given medical treatment rather than the starring role in a horror story.

Even if the *Ruminococcus* enterotype does not exist, there is no doubt that these bacteria are present in our gut. So it is useful that we now know more about them — and about Dracula and red pee. Our bacteria-free laboratory mice have trouble forming haem, and so it stands to reason that bacteria are somehow important in this process.

Now we are more familiar with the tiny world of the microbes in our gut. Their genes represent a huge pool of borrowed abilities. They help us digest our food, and they produce vitamins and other useful substances. We are just beginning to recognise enterotype commonalities and to search for patterns in them. And we do this for one reason: 100 trillion tiny creatures reside in our tummies, and that cannot but have an effect on us. So let's now go one step further, and explore the palpable effects they have on us. Let's take a closer look at how these gut bacteria affect our metabolism, and examine which ones do us good, and which ones do us harm.

The Role of the Gut Flora

Sometimes we tell fibs to our children. Because they are so nice, like the one about the man with the big white beard who arrives once a year on his tuned-up reindeer sleigh piled high with children's presents. Or the one about the Easter bunny hiding chocolate eggs in the garden. Sometimes, we don't even realise when we are not telling the truth. Like when we encourage a toddler to eat up: 'One spoon for Daddy, one spoon for Mummy, one for Granny, and one for Granddad ...' If we wanted to encourage junior in a scientifically correct way, we would have to say, 'One spoon for you, baby. A small part of the next spoon for your *Bacteroides* bacteria. An equally small part for your *Prevotella*. And a teeny-weeny bit for a few other micro-organisms waiting in your tummy to be fed.' We might well send a friendly vote of thanks down to the micro-colleagues enjoying the meal in baby's belly. After all, *Bacteroides* and company work hard to help keep baby well fed. And not only in infancy. Adults, too, receive nutrition back from their bacterial gut-dwellers, morsel by morsel. Gut bacteria process food that we cannot break down unaided, and share the results with us.

The idea that the bacteria in our gut might influence our overall metabolism, and therefore our weight, is only a couple of years old. The basic concept is that bacteria do not steal anything from us when they share our food in this way. Very few gut bacteria reside in the small intestine, where we break down our food for ourselves and absorb the nutrients from it. The highest concentration of bacteria is found where the digestive process is almost finished, and all that remains is for the undigested remnants to be transported away. The further you travel from the small intestine towards the final exit from the gut, the more bacteria you will find per square centimetre of gut membrane. It is our gut that makes sure this remains so. If the equilibrium is disturbed and large numbers of over-confident bacteria migrate to the small intestine, we have a case of what doctors call 'bacterial overgrowth'. This relatively unexplored condition causes symptoms that can include severe bloating, abdominal pain, joint pain, and gastrointestinal infections, as well as nutrient deficiencies and anaemia.

In ruminants such as cows, this construction design is reversed. These large animals are well known for their ability to survive by eating only grass and a few other plants. They are far from being puny vegans, so what is their secret? Cows keep their bacteria right at the top of their digestive tract. They don't even bother trying to digest their food themselves first, but pass the complex plant carbohydrates straight on to *Bacteroides* and Co. Their microbes turn them into an easily digested feast for the cow.

It can be practical to keep your bacteria so close to the beginning of your digestive tract. Bacteria are rich in protein — so, from a food point of view, they are tiny little steaks. When they have finished their life's work in the cow's stomach, they slip further down the system, where they are digested. They are a large source of protein for the cow — tiny microbial steaks, bred by cows themselves. Our bacteria are too far down the system to provide this practical steakhouse service; instead, we pass them out of our gut undigested.

Rodents keep their microbes as far down the system as we do, but are more loath to waste the bacterial protein they contain. Their simple solution is to eat their own faeces. We don't do that, preferring to buy meat or tofu at the supermarket to compensate for the fact that we are unable to process protein-rich bacteria in our large intestine. However, we still benefit from their work, even if we don't digest them. Bacteria produce nutrients that are so tiny we can absorb them directly into the cells of our gut.

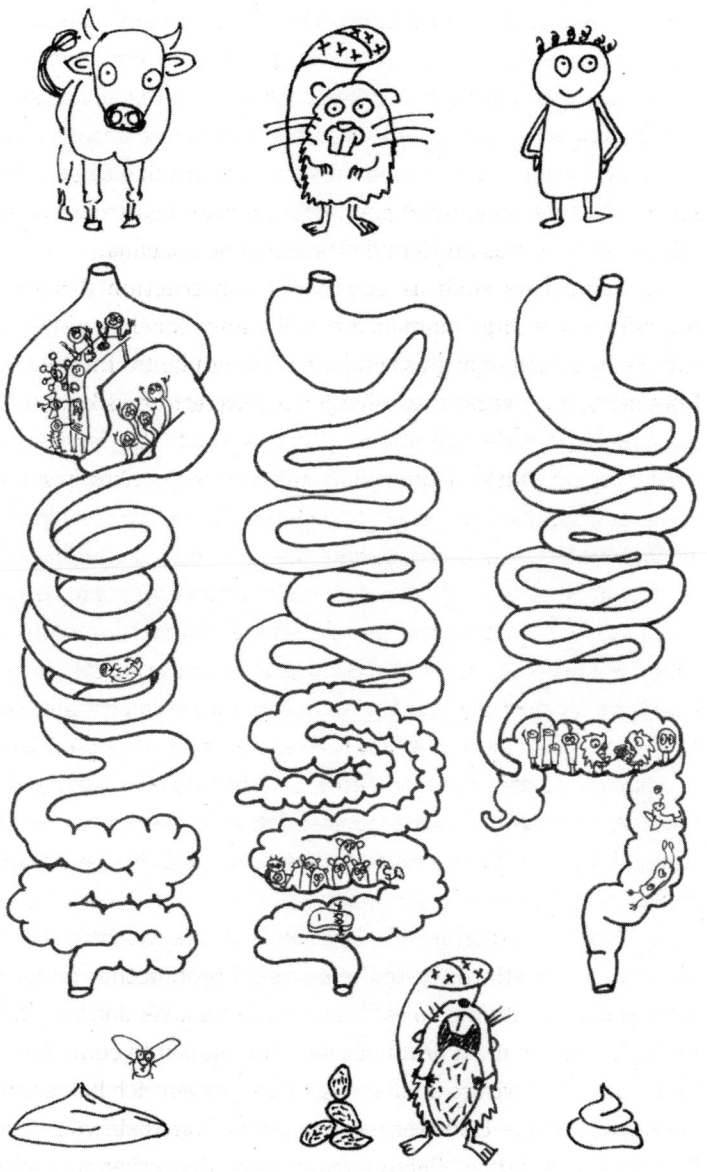

They can also perform this service outside the gut. Yoghurt is nothing other than milk that has been pre-digested by bacteria. Much of the sugar in the milk (lactose) has already been broken down and transformed into lactic acid (lactate) and smaller sugar molecules. That is why yoghurt is both sweeter and sourer than milk. The newly formed acid has another effect on the milk: it causes the milk protein to curdle, giving the yoghurt its characteristic thick consistency. Pre-digested milk (yoghurt) saves our body some work — we just have to finish off what the bacteria started.

It is an especially good idea to employ bacteria that manufacture healthy end-products to pre-digest our food. Mindful yoghurt manufacturers often use bacteria that produce more 'dextrorotatory' (right-turning) than 'levorotatory' (left-turning) lactic acids. Molecules of the two kinds of lactic acid are mirror images of each other. Feeding the human digestive system with levorotatory lactic acid molecules is like giving a left-handed pair of scissors to a right-handed person: they're hard to handle. That is why it is a good idea to pick yoghurt from the supermarket shelves that says 'contains mainly dextrorotatory (or right-turning) lactic acid' on the pot.

Bacteria do more than just break down our food. They also produce completely new substances. Fresh cabbage, for example, is less rich in vitamins than the sauerkraut it can be turned into — those extra vitamins are made by bacteria. Bacteria and fungi are responsible for the taste, creamy consistency and holes in cheese. Bologna sausage (baloney) and salami are often made with 'starter cultures' — that's butchers' code for 'We daren't tell it to you straight, but it is the bacteria (mainly *Staphylococcus*) that makes it so tasty.' Lovers of wine or vodka appreciate the metabolic end-product of yeasts — known as alcohol. The work of these micro-organisms does not end in the wine barrel. Almost none of what a wine taster will tell you is actually to be found in the bottle. The wine's 'bouquet', for example, develops so late because bacteria need time to do their work. They sit waiting at the back of the tongue, where the process of transforming what we eat or drink begins. The substances they release during that process create the aftertaste so appreciated by the wine lover. And each connoisseur will experience a slightly different taste — depending on the population of bacteria on their tongue. Still, it's nice to get such an enthusiastic reaction to the presence of these much-maligned microbes.

The bacteria population of our mouth is only about one ten-thousandth of the number that live in the gut, yet we still taste the results of their work. Our digestive tract should be grateful for the fact that it has such a large population with such a wide range of skills. While simple glucose and fructose are easy digested, many people's guts start to flag when it comes to lactose, the sugar contained in milk. Their owners then suffer from a lactose intolerance. Complex plant carbohydrates would flummox a gut if it were expected to have at the ready every enzyme needed to break them down. Our microbes are experts in dealing with these substances. We provide them with somewhere to live, and the undigested remains of our food — and they keep busy dealing with the stuff that's too complicated for us to do.

In the industrialised world, about 90 per cent of our nutrition comes from what we eat, and we are fed about 10 per cent by our bacteria. So, after nine lunches, meal number ten is on the house, so to speak. Feeding adults is the main occupation of some of our bacteria. That does not mean that it makes no difference what we eat — far from it. And the kind of bacteria that are feeding us also makes a big difference. In other words, if we are concerned about our weight, we need to think about more than just the big, fat calories we consume, and to remember that our bacteria are at the dinner table with us.

How might bacteria make us fat? Three theories

1.

The gut's flora include too many 'chubby bacteria'. Chubby bacteria are efficient at breaking down carbohydrates, but if the number of chubby bacteria gets out of hand, we have a problem. Skinny mice excrete a certain quantity of indigestible calories, while their overweight peers excrete significantly fewer. Their chubby bacteria extract every last smidgen of energy from the same amount of food, and cheerfully feed it to Mr or Ms Mouse. For humans, this can mean that some people pile on weight even though they don't eat more than others — it could be that their gut flora are extracting more energy from the food they eat.

How is this possible? Bacteria are able to make various fatty acids out of indigestible carbohydrates: vegetable-loving bacteria tend to manufacture fatty acids for the gut and the liver; others produce fatty acids that feed the rest of the body. Thus, a banana is less likely to make you fat than half a chocolate bar containing the same number of calories. That is because plant carbohydrates are more likely to attract the attention of bacteria that provide fatty acids to local customers like the liver. Chocolate, on the other hand, is more likely to attract the attention of the full-body feeders.

Studies carried out on obese subjects show that they have less overall diversity in their gut flora, and that certain groups of bacteria prevail — primarily those that metabolise carbohydrates. To succeed in becoming obese, however, a few other conditions must be met. Experiments with laboratory mice showed that some weighed 60 per cent more at the end than at the beginning of the trial. Bacterial 'feeders' cannot cause so much weight gain alone. This prompted scientists to consider another marker for extreme weight-gain: inflammation.

2.

Patients who have metabolism problems, such as obesity, diabetes, and high blood-lipid levels, usually have slightly increased levels of inflammation markers in their blood, too. They are not so high that they require treatment, as they would if they were caused by an inflammation wound or septicaemia. That's why doctors call this a 'subclinical inflammation'. If there is anything that knows a thing or two about inflammation, it's bacteria. They have a signalling substance on their surface that tells the body when to get inflammation.

When we get injured, this reaction is useful. Inflammation reactions flush out bacteria, or attack them. As long as bacteria remain in their cosy mucous-membrane home in the gut, this signalling substance goes unnoticed. But when bacteria appear in disadvantageous combinations, or when their host eats an overly fatty diet, too many of them can find their way into the bloodstream. The body then slips into low-key inflammation mode. From the evolutionary point of view, it's worth paying that price to build up fat reserves for leaner times.

Bacterial signalling substances can also latch onto other organs and affect the metabolism in that way. In rodents and humans, they dock onto the liver or the fatty tissue itself, and encourage the deposition of more fat. They also have an interesting effect on the thyroid gland. Bacterial inflammation hinder its function, causing it to produce fewer thyroid hormones, slowing the rate at which the body burns fat.

Unlike acute inflammation, which cause weight loss or even emaciation, subclinical inflammation causes weight gain. Bacteria are not the only possible cause of subclinical inflammation — hormone imbalances, too much oestrogen, lack of vitamin D, or too much gluten-rich food have all been observed to have a similar effect.

3.

Now, brace yourself for a crazy idea! A hypothesis postulated in 2013 suggests that gut bacteria can affect their host's appetite. Roughly speaking, the theory is that late-night cravings for chocolate-covered toffees followed by an entire bag of party pretzels do not originate in the organ that calculates our tax returns. Not our brains, but our guts are the home of gangs of bacteria that crave hamburgers after three days on a diet. Somehow they manage to pass on that message in a very persuasive way, because we find it almost impossible to deny them any wish.

To understand this hypothesis, we have to try to get our head round the issue of food. Faced with a choice between two different dishes, we make a decision based on what we happen to fancy at that moment. The amount of our chosen dish that we actually eat is controlled by the feeling of satiety. In theory, bacteria have ways and means of influencing both those things. Again, we can currently only conjecture that they also have a say in our appetite. But in evolutionary terms it is not such a silly idea. What and how much we eat can be a matter of life and death for them. In three million years of co-evolution, even simple bacteria have had time to adapt optimally to life with their human hosts.

If you want to trigger a craving for specific foods, you have to gain access to the brain. That is no mean feat. The brain is wrapped in a sturdy membrane called the meninges. Even more impenetrable than the meninges are the coats that surround all the blood vessels passing through the brain. The only things that can make it through this tangled mess are pure sugar, minerals, and anything that is as small and fat-soluble as a neural transmitter. Nicotine, for example, makes it through to the brain, where it triggers reward signals or a feeling of relaxed alertness.

Bacteria can produce particles that are small enough to make it through the coating of the blood vessels, into the brain.

Examples include *tyrosine* and *tryptophan*. These two amino acids are converted into dopamine and *serotonin* in the cells of the brain. Dopamine? Wasn't that famously associated with the brain's reward system? And *serotonin*? That sounds familiar, too, doesn't it? Lack of it causes depression. It can make us feel contented and sleepy. Just think about last year's Christmas dinner. Did you end up dozing on the couch after enjoying it, feeling lethargic and sleepy, but also contented?

So the theory is this: our bacteria reward us when we send them a decent delivery of food. It feels pleasant, and whets our appetite for the next meal. They do this not only directly by means of the substances they produce, but also by cranking up the body's production of certain transmitters. The same principle applies to the feeling of satiety.

Several studies have shown that our satiety-signal transmitters increase considerably when we eat the food that our bacteria prefer. And what our bacteria prefer is food that reaches the large intestine undigested, where they can then gobble it up. Surprisingly enough, those foods do not include pasta and white bread. For more on this, turn to page 237 (the section on *prebiotics*).

The feeling of satiety generally comes from two sides: from the brain, and from the rest of the body. A lot can go wrong here. In obese people, the gene that codes for satiety may be defective, and such people simply do not get that full feeling after eating. According to the 'selfish brain' theory, the brain does not receive enough of the energy eaten as food, and so decides that it is still hungry. But it is not only our body tissue and our grey matter that depend on the food we eat — our microbes also need to be fed. They may seem small and insignificant by comparison with our body size — accounting for just 2 kilos of our body weight. So what right do they have to butt in?

When we consider the range of functions carried out by our gut flora, it is not surprising that these microbes are also able to express their wishes. They are, after all, the immune system's most important trainers, digestion assistants, producers of vitamins, and experts in detoxifying mouldy bread or medical drugs. The list is much longer, of course, but the message should be clear: they should be able to have a say in whether we feel full after eating.

We do not yet know whether different bacteria express different desires. When we give up sweets, we eventually stop missing them so badly at some point. Is that because the gummy bear and chocolate lobby has been starved out? We can only speculate.

The important thing is not to reduce the human body to a two-dimensional cause-and-effect machine. The brain, the rest of the body, bacteria, and the elements in our food all interact with each other in four dimensions. Striving to understand all these axes is surely the best way to improve our knowledge. However, we can more easily tinker with bacteria than with our brain or our genes — and that is what makes microbes so fascinating. The nutrition we receive from our bacteria is not only important for fighting flab, it also affects the levels of fats such as cholesterol in our blood, for example. This realisation could be quite highly charged: obesity and high cholesterol levels are closely connected with the greatest health issues of our time: hypertension, arteriosclerosis, and diabetes.

Cholesterol and gut bacteria

The connection between bacteria and cholesterol was first discovered in the 1970s. American scientists studying Maasai warriors in Africa had been surprised to find that the levels of cholesterol in their blood were low, despite a diet consisting almost entirely of meat and milk. The excessive amount of animal fat did not cause high blood-lipid levels. The scientists suspected a mysterious substance in the milk they drank of causing their cholesterol levels to remain low.

They then set about doing all they could to find that mysterious ingredient. They tested cow's milk, camel's milk, and even rat's milk. Sometimes they managed to reduce cholesterol levels; sometimes not. These results told the scientists nothing. In another experiment, they gave the Maasai warriors a vegetable-based milk-replacement product (Coffee-Mate) with high levels of cholesterol added to it. Still, the subjects' blood cholesterol levels did not rise. The scientists saw that their theory about a mysterious milk component had been disproven.

They had meticulously noted that the Maasai often drank 'curdled' milk. But no one considered the fact that certain bacteria are required to curdle milk — if they had, it would have explained the results of their 'Coffee-Mate experiment'. Bacteria that have already settled in the gut can continue to live there even when milk is replaced with a plant-based creamer enriched with cholesterol. Instead, even when the Maasai's cholesterol levels were seen to sink by 18 per cent whenever they drank 'curdled' rather than fresh milk, scientists continued to search for the mysterious milk substance. This was blind dedication that brought no results.

These studies of the Maasai would not live up to modern scientific expectations. The test groups were too small, and the Maasai spend about 13 hours a day walking and one month a year fasting. They simply could not be compared to meat-eating Europeans. However, the results of these studies were rediscovered decades later by researchers who were now aware of the importance of bacteria. Cholesterol-lowering bacteria? Why not test that in the lab? Take a flask of nutrient broth warmed to a balmy 37°C, add cholesterol and some bacteria — *et voilà*. The bacteria they used was *Lactobacillus fermentus*, and the cholesterol they had added was ... gone! At least, a considerable proportion of it was.

Experiments can have very different results, depending whether they are carried out in glass flasks or inside *opisthokonts*.

It's like an emotional rollercoaster ride for me when I read in scientific papers sentences like 'The bacterium L. plantarum Lp91 can significantly lower high cholesterol and other blood lipid levels, increase 'good' HDL cholesterol and lead to significantly lower rates of arteriosclerosis, *as could be shown in 112 Syrian golden hamsters.*' I have never been so disappointed by Syrian golden hamsters. Experiments on animals are a good way to begin tests on living systems. If the sentence had ended 'as could be shown in 112 obese Americans', the whole thing would be a lot more impressive.

But a result like this can still be worth a lot. Studies on mice, rats, and pigs yielded such promising results for some bacteria that scientists considered it reasonable to begin human testing. The subjects were regularly administered certain bacteria, and after a period of time their cholesterol levels were measured. The bacteria species, the quantities, the duration, and the way they were administered all varied. Some results were positive; others were not. Also, no one really knew whether sufficient numbers of the administered bacteria even survived their bath in the acid juices of the stomach long enough to have an effect on blood-cholesterol levels.

Really valuable studies began about twenty years ago. For one experiment in 2011, 114 Canadians ate specially produced yoghurt twice a day. The bacteria added to the yoghurt was *Lactobacillus reuteri* — in a form particularly resistant to digestion. Within six weeks, their levels of 'bad' LDL cholesterol sank by 8.91 per cent. That's about half the improvement attained by taking a mild anti-cholesterol drug — but without the side effects. Studies using other types of bacteria lowered cholesterol levels by as much as 11 to 30 per cent. Follow-up research still needs to be carried out to verify these promising indications.

There are several hundred bacteria candidates that might be tested in future. To sift out the less likely ones, we have to ask questions like: What abilities does such a bacterium need to have? Or, rather, what genes does it need to have? The most likely candidates we know of today are BSH genes. BSH stands for 'bile salt hydroxylase'. Bacteria with these genes can convert bile salts. But what do bile salts have to do with cholesterol? The answer is in the name. Cholesterol comes from the Greek words *'chole'* — bile, and *'stereos'* — solid. Cholesterol was first discovered in gallstones. Bile, which is stored in the gall bladder, is the body's transport medium for fats and cholesterol. BSH allows bacteria to alter bile to make it work less efficiently. The cholesterol and fat dissolved in bile can then no longer be absorbed by the body and they end up, to put it bluntly, down the toilet. This mechanism is useful for bacteria, because it weakens the effect of bile that might attack their cell membranes. This protects them on their long journey to their final destination — the large intestine. Bacteria also have a few other mechanisms for dealing with cholesterol: they can absorb it directly and incorporate it in their cell walls; they can convert it into a new substance; or they can manipulate organs that produce cholesterol. Most cholesterol is produced in the liver and the gut, where tiny messenger substances manufactured by the bacteria can partly control those processes.

But we need to take a step back and ask cautiously: Does the body always want to get rid of its cholesterol? It produces between 75 and 90 per cent of our cholesterol itself. And that takes a lot of work! One-sided media reporting has given cholesterol a bad name, making people believe it is only evil. That is quite wrong. *Too much* cholesterol is not such a good thing, but neither is *too little*. If it weren't for cholesterol, we would have no sex hormones, no vitamin D, and a plethora of unstable cells. Fat and cholesterol are not only a problem for Granny because of her weakness for cream cakes or sausages. They are a problem for every one of us. Studies have shown a connection between too little cholesterol and memory problems, depression, and aggressive behaviour.

Cholesterol is also the miraculous base material for all sorts of structures. Too much of it is indeed harmful — it's all about finding the right balance. Our bacteria would not be our bacteria if they didn't have the ability to help us achieve that. Some bacteria produce more of a substance called *propionate*, which inhibits the production of cholesterol. Others produce more acetate, which promotes the production of cholesterol.

Who would have thought it? A chapter that began with bright little spots of bacteria ending with concepts like appetite and satiety, or substances like cholesterol? To summarise: bacteria help to feed us, make some foods more digestible, and produce their own substances. Some scientists now support the theory that our gut microbiota can be considered an organ. Just like the other organs in our body, it has an origin, develops along with us, is made up of a load of cells, and is in constant contact with its fellow organs.

The Bad Guys:
harmful bacteria and parasites

There are good guys and bad guys in the world — and the same goes for the world of our microbes. One thing unites most of the bad guys: they only want what's best ... for themselves.

Salmonellae in hats

Even the most courageous of cooks sometimes feel a pang of primal fear while beating eggs — fear of the raw threat posed by salmonella! Everyone knows someone who has endured devastating diarrhoea and venomous vomiting after nibbling on a bit of raw cake mix, or eating chicken that was not quite done.

Salmonella bacteria can get into our food in unexpected ways. Sometimes, globalisation helps them find a home in our chicken meat or eggs. This is how it can happen: the cheapest source of feed grain for chickens is Africa. So, in Europe, we import it to feed the fowl in our poultry farms. However, there are more wild tortoises and lizards wandering around in Africa than Germany. Salmonella bacteria travel to our climes along with the chicken feed. How so? Well, they are part of the normal gut flora of reptiles. While the African farmer is working in the fields, a tortoise might merrily be doing its business in a sack of grain destined for Germany. After a relaxing sea voyage around the coasts of the European continent, the grain, along with its stowaway tortoise-poo bacteria, ends up in a German poultry farm, where it is eaten by a hungry chicken. Salmonella bacteria are not part of a chicken's natural gut flora, but they are a common pathogen.

Once inside the bird's gut, the salmonellae can multiply, and they are eventually excreted. Since chickens have only one hole for all export goods, the egg cannot avoid coming into contact with salmonellae in the bird's faeces. The bacteria are then usually found only on the shells of eggs — they only get inside when the shell is cracked.

But what about salmonellae in chicken meat? How do they get there? That's a rather unsavoury story. Cheaply fed chickens are usually sent to large industrial slaughterhouses to meet their maker. Once they have been slaughtered and beheaded, they are dunked in huge tanks of water. Those tanks are like a wellness spa for salmonella bacteria, complete with colonic irrigation service for the chickens. In a slaughterhouse dispatching 200,000 birds a day, one batch of cheap-feed chickens is enough to give the gift of salmonella to all the other birds in the bath. The chickens then end up in the freezers of discount supermarkets. If they are roasted or grilled at high-enough temperatures, the salmonella germs are soon killed off, and are no longer able to cause problems for anyone.

Properly cooked meat is normally not the reason for most salmonella infections. The problems begin when the frozen chicken is left to thaw in the kitchen sink or colander. Freezing and thawing does bacteria little harm. The huge library of bacteria in our laboratory includes germs collected from patients, which easily survive temperatures of -80°C and live merrily on when thawed. Heat is their nemesis — even just ten minutes' exposure to a temperature of 75°C is enough to see off all salmonella bacteria. That's why a carefully roasted chicken is not usually the culprit, but rather the lettuce leaves for the side salad, left briefly to soak in the same kitchen sink.

We come into regular contact with the gut flora of the livestock we keep, but we only notice it when they happen to include unfamiliar, diarrhoea-causing bacteria. The rest is routine, so to speak, and, after all, we have to get our bacteria from somewhere. Sticking to organic country eggs from chickens fed with homegrown grain is usually a reliable way to avoid dangerous bacteria — unless the farmer himself eats cheap chicken from a discount supermarket.

If our Sunday roast chicken is not roasted enough, we may end up eating not just chicken muscle cells, but a few salmonella cells, too. It takes between 10,000 and one million of these single-celled creatures to put us out of action. A million of these bacteria take up about one-fifth as much space as a grain of salt. So how does an army of such tiny soldiers manage to move a colossus like us, with the volume of about 600,000,000 grains of salt, inexorably towards the toilet? It's as if one hair of Obama's head were to rule over the entire population of America.

Salmonella bacteria double in number much more quickly than politicians' hairs — that's point one. As soon as the temperature rises above 10°C, salmonellae come out of hibernation and get busy breeding. They have delicate little arms that enable them to swim around until they find a place to attach to the gut. Once there, they invade our cells, which become infected and pump large quantities of fluid into the gut in an attempt to flush out the pathogen as quickly as possible.

It can take a few hours to a few days from accidental ingestion to watery flush-out. A self-induced colonic irrigation like this usually works well, unless the victim is too young, too old, or too frail. Antibiotics would do more harm than good here. Despite this natural cure, it is better for the gut to flatly refuse entry to salmonella, however rude that may seem. After a visit to the toilet or a retching session into a sick bag, you should not take them by the hand and show them what life is like in the outside world. They should be shown the cold shoulder by washing with very hot water and soap to let them know: it's not you, it's me — I just can't do with your clinging personality.

Salmonellae are the most common type of bad guys we take in with our food. They are not only found in poultry products, but they are one of salmonella's favourite hang-outs. They come in several varieties. When we receive stool samples from patients at our laboratory, we can test them by exposing them to different antibodies. When an antibody bonds with the salmonellae, they clump together into blobs that are big enough to see with the naked eye.

When that happens, we can say that the antibody to vomit-inducing salmonella XY reacts strongly, so this must be vomit-inducing salmonella XY. This is the same as the mechanism in our own bodies. Our immune system meets a couple of new salmonellae, and says, 'Hmmm, I'm sure I must have a hat that fits them somewhere in my collection.' It then rummages around in its wardrobes to find the right hat, adjusts it a little to create the perfect fit, then commissions a hatter to make headgear for a million salmonella germs. When all the salmonella bacteria are wearing their new hats, they no longer look dangerous, but rather ridiculous. They are weighed down so much by the millinery that they are too heavy to swim around attacking anything. In this way, the test antibodies in the lab can be seen as a small selection of different hats. When the hat fits, the heavily sombreroed bacteria collapse into clumps and — depending on the hat — we can say which type of salmonella was in the stool sample.

For those who do not necessarily want to send their immune system searching for hats, and those who are no great fans of diarrhoea and vomiting, there are a few simple rules to follow.

Rule number one: always wash anything that comes into contact with raw meat or eggshells thoroughly with hot water — chopping boards, hands, cutlery, kitchen sponges, and colanders, for example.

Rule number two: whenever possible, make sure that meat and egg-based foods are cooked through. Of course, that doesn't mean you have to interrupt your romantic dinner to stick the tiramisu in the microwave. If you are planning a dish of that sort, make sure you always use good-quality, fresh eggs, and always store them at a temperature of less than 10°C.

Rule number three: think beyond the kitchen. Anyone who has had to rush to the toilet after feeding their pet iguana and then themselves (without washing their hands properly) will remember my words: salmonella bacteria are part of the normal gut flora of reptiles.

Helicobacter: humanity's first 'pet'

Thor Heyerdahl was a phlegmatic man with strongly held views. He observed ocean currents and winds. He was interested in ancient fishhooks and clothing made of tree bark. All this convinced him that the Polynesian islands were first colonised by seafarers from South America and South-East Asia. He theorised that they could have used currents to help them reach the islands on rafts. At the time, no one thought it possible that a simple raft could survive an 8,000-kilometre journey across the Pacific. Thor Heyerdahl was not a man to waste time trying to convince doubters with theoretical arguments. He went to South America, built a primitive raft out of balsawood logs, grabbed a couple of coconuts and tins of pineapple, and set out for Polynesia. After four months, he could safely say: 'Yes, it is possible!'

Thirty years later, another scientist set off on an equally exciting expedition. But his journey did not take him across oceans; it took him to a small laboratory with neon strip lights on the ceiling. There, Barry Marshall picked up a petri dish of liquid, placed it to his lips, and bravely swallowed its contents. His colleague observed him with interest as he did so. After a few days, Barry Marshall developed a stomach inflammation and could proudly say 'Yes, it is possible!'

Another thirty years went by before scientists in Berlin and Ireland made a connection between the researches of these two very different pioneers. Marshall's stomach bug was destined to provide information about the first colonisation of Polynesia. This time, no one sailed an ocean, and no one drank a lab culture. Researchers asked a few desert-dwelling aboriginals in Australia and highland tribesmen from New Guinea for a sample of their stomach contents.

This is a story about disproving paradigms, dedication to research, a tiny creature with a propeller, and a big, hungry cat.

The bacteria *Helicobacter pylori* lives in the stomachs of at least half of humankind. This insight is relatively new, and was initially ridiculed. Why should an organism live in such an inhospitable environment — a cave full of acid and enzymes bent on breaking them down? It takes more than that to discourage *Helicobacter pylori*. This bacterium has developed two strategies that enable it to cope excellently with such a harsh environment.

First, the products of its metabolism are so alkaline that that they can neutralise any acid in its immediate vicinity. Second, it burrows beneath the mucous membrane that protects the stomach from digesting itself with its acidic juices. This membrane usually has a gelatinous consistency, but *Helicobacter* is able to liquefy it so that it can swim more easily through the mucoid lining. It has long threads of proteins it uses as flagella to whizz around.

Marshall and Warren believed *Helicobacter* caused stomach inflammations (gastritis) and gastric ulcers. The prevailing scientific opinion at the time was that those stomach problems were psychosomatic in origin (as a result of stress, for example), or caused when the stomach secreted too much acid. Marshall and Warren had to both counter the preconception that nothing could survive in the acid environment of the stomach and prove that a tiny bacterium could cause diseases that were not traditional bacterial infections. Until that time, it was believed that bacteria could only infect wounds, and cause fevers and colds.

After the otherwise healthy Marshall deliberately swallowed *Helicobacter* bacteria and gave himself gastritis, which cleared up after a course of antibiotics, it took another ten years before the scientific world accepted his discovery. Today, it is standard practice to test patients with stomach complaints for the presence of these bacteria. The patient is given a fluid to drink, and if *Helicobacter* are present in the stomach, they break down the ingredients of the fluid, and a resulting marked, odourless gas can be identified in the breath of the patient. Drink, wait, breathe — a relatively simple test.

What the two scientists did not realise was that they had discovered not only a cause of illness, but also one of humankind's oldest 'pets'. *Helicobacter* bacteria have been living inside human beings for more than 50,000 years, and have evolved in parallel with us. When our ancestors began to migrate around the world, *Helicobacter* went along for the ride and founded new populations, just like those human pioneers. Today, three African, two Asian, and one European type of this bacterium have been identified. The further the population groups spread from each other in both space and time, the greater the difference between their stomach bacteria.

The slave trade transported the African types to America. In northern India, the Buddhist and Muslim populations have different strains in their stomachs. Families in industrialised nations often have their own family strain of *Helicobacter*, while people living in societies with more contact between individuals — in parts of Africa, for example — have communal *Helicobacter* strains.

One out of three Europeans is a carrier for *H. pylori*. Not everyone who carries it is doomed to develop stomach problems, but most people who do have *H. pylori* to thank for their woes. This is because some *Helicobacter* bacteria are more dangerous than others.

Two factors are known to be responsible for the more virulent version. One is called 'cagA', and it enables the bacteria to inject certain substances into our cells like a tiny syringe. The second factor is called 'VacA'. It needles the cells of the stomach continuously, causing them damage that is eventually fatal to the cell. There is a much higher probability of developing stomach problems if the *Helicobacter* microbes in the stomach possess the injecting-syringe or cell-damaging gene. If those genes are not present, *Helicobacter* is much less harmful as it swims around in the stomach.

Although they share many similarities, each *Helicobacter* bacterium is as unique as the person carrying it. These bacteria adapt to their host, and change as she changes. Scientists can make use of this fact to trace who infected whom with the germ. Big cats have their own feline *Helicobacter*. Its name is *Helicobacter acinonychis*. The fact that it bears such a resemblance to human *Helicobacter* provokes an obvious question: who was eating whom in prehistoric times? Was it a case of man eating tiger, or tiger eating man?

Genetic analysis shows that the genes that were deactivated in the feline version of the bacterium were mainly those that would otherwise enable it to latch onto the cells of the human stomach — but the reverse is not the case. When devouring a prehistoric person, large cats must also have devoured their stomach bacteria. These are not killed by even the sharpest of tiger teeth, so *Helicobacter* colonised the stomach of the predator and its descendants. A tiny bit of redress, at least.

But is *Helicobacter* good or bad?

Helicobacter is bad

By infiltrating our stomach's mucous membrane and swimming around in it, *Helicobacter* weakens this protective barrier. As a result, the aggressive acids in our stomach digest not only our food, but a little bit of our own stomach, too. If the bacteria also possess the injection-syringe or cell-damaging gene, our stomach cells have little hope. About one-fifth of people who harbour this bacterium develop tiny lesions in their stomach wall. Two-thirds of stomach ulcers and almost all ulcers in the small intestine are caused by a *Helicobacter pylori* infection. If these microbes are wiped out with antibiotics, the patient's stomach problems disappear. Even better, a new product could soon provide an alternative to antibiotics: *sulforaphane* — contained in broccoli and similar vegetables. This substance is able to block the enzyme that *Helicobacter* uses to neutralise gastric acid. Those who would like to try it as an alternative to antibiotics should make sure that they use very high-quality broccoli, and that they consult their doctor after two weeks, to test whether their *Helicobacter* population has really disappeared.

Constant irritation is never a good thing. We are all familiar with itchy insect bites — at some point we can no longer resist scratching them to make the itching stop, even though we know we will end up with a bleeding wound. Something similar happens with the cells of the stomach. A chronic inflammation means the cells are permanently irritated, until they finally give up the ghost and break down. In older people, this can also be a cause of appetite loss.

The stomach has a battery of stem cells that constantly replace lost cells. If these replacement manufacturers are overworked, they may begin to make mistakes. Cancer cells are the result. Statistically, this does not look too serious: around 1 per cent of *Helicobacter* carriers develop stomach cancer. But if you bear in mind that half of humanity harbour these bacteria in their stomachs, 1 per cent turns out to be a pretty big number. The probability of developing stomach cancer without the presence of *Helicobacter* is about forty times less than with it.

In 2005, Marshall and Warren received a Nobel Prize for their discovery of the connection between *Helicobacter pylori* and gastritis, stomach ulcers, and cancer. The journey from a bacteria cocktail in Perth to a celebratory cocktail in Stockholm took twenty years.

It took even longer for the connection between *Helicobacter* and Parkinson's disease to be realised. Although doctors had known since the 1960s that patients with Parkinson's have an increased incidence of stomach problems, they did not know the nature of the connection between sore stomachs and trembling hands. It took a study of different population groups on the Pacific island of Guam to throw light on the subject.

In some parts of the island, there was an astonishingly high incidence of Parkinson-like symptoms among the population. Those affected suffered from trembling hands, facial paralysis, and motor problems. Researchers realised that the symptoms were most common in areas where people's diets included cycad seeds. These seeds contain neurotoxins — substances that damage the nerves. *Helicobacter pylori* can produce an almost identical substance. When laboratory mice were fed with an extract of the bacteria without being infected with the living bacterium itself, they displayed very similar symptoms to the cycad-eating Guamanians. Once again, not every *Helicobacter* bacterium produces this substance, but if they do, it is not good news.

In summary, it can be said that *Helicobacter* is able to manipulate our protective barriers, irritate and destroy our cells, and manufacture toxins — and damage our entire body by doing so. So how have our vulnerable bodies been able to survive many millennia of infection with this bad bacterium? Why have these bacteria been so widely tolerated by our immune system for so long?

Helicobacter is good

A large-scale study of *Helicobacter* and its effects reached the following conclusion: the bacterium, especially that much-feared strain with the injection-syringe gene, can also interact with the body in beneficial ways. After more than twelve years of observing over 10,000 subjects, it was concluded that carriers of this type of *Helicobacter* do have an increased risk of stomach cancer, but that their risk of dying of lung cancer or a stroke was much lower. In fact, it was only half that of other subjects in the study.

Even before this study was conducted, scientists had already suspected that a bacterium that has been tolerated for so long cannot be only bad. They had shown in experiments with mice that *Helicobacter* provides a reliable protection against childhood asthma. When the mice were given antibiotics, this protection disappeared and the infant rodents stood a chance of developing asthma again. When the bacterium was given to adult mice, the protective effect was still present, but much less pronounced. You might say that mice are not people, but this observation fits very well with the general trend noticed in industrialised countries in particular: rates of asthma, allergies, diabetes, and neurodermatitis have risen, as rates of *Helicobacter* have fallen. This observation does not constitute proof that *Helicobacter* is the sole protection against asthma — but it may be part of the overall picture.

The theory suggested to explain this is that this bacterium teaches our immune system to stay cool. *Helicobacter* latches onto our stomach cells, and by doing so causes large numbers of regulatory T-cells to be produced. These are immune cells whose job is to place a calming hand on the shoulder of the immune system when it flies off the handle like a drunk in a crowded nightclub, saying, 'Let me deal with this, mate.' As the name implies, they regulate the immune system's reactions.

While the irate immune system is still shouting, 'Get out of my lungs, you bloody pollen-thing!' and showing its readiness to fight by giving us swollen red eyes and a runny nose, the regulatory T-cells say, 'Oh, come on, immune system, that was a bit of an overreaction. The pollen-thing is only looking for a flower to pollinate. It's just landed here by mistake. That's more of a problem for the pollen grain than us. It will never find its flower now.' The more of these sensible cells there are at work, the more chilled the immune system will be.

When *Helicobacter* causes a particularly large number of these cells to be produced by one mouse, another mouse's asthma can

be improved simply transplanting those cells to it. That must be an easier solution than training a mouse to use a tiny little inhaler!

The incidence of eczema seems to be reduced by about a third in people who harbour *Helicobacter pylori*. Increases in inflammatory gut disease, autoimmune problems, and chronic inflammations may also be a modern trend caused by the fact that we often unwittingly wipe out something that has protected us for millennia.

Helicobacter is good and bad

Helicobacter pylori are bacteria with many capabilities. They can't be labelled simply as good or bad. It always depends on what exactly the bacterium does inside us. Is it manufacturing dangerous toxins, or is it interacting with our body to protect it in some way? Are our cells constantly irritated, or can we produce enough gastric mucous for both its needs and our own? What part is played by agents that irritate the stomach's mucous membrane, such as painkillers, cigarette smoke, alcohol, coffee, and stress? Is it a combination of all these that is responsible for stomach ulcers, because our little pets don't like them?

The World Health Organisation recommends that people with stomach problems should get rid of the potential culprits. If stomach cancer, certain lymphomas, or Parkinson's disease run in the family, it is also a good idea to offload *Helicobacter*.

Thor Heyerdahl died in Italy in 2003 at the age of 88. Had he lived just a couple of years longer, he would have seen his theory about the colonisation of Polynesia confirmed by studies of *Helicobacter* strains. Asian strains of *Helicobacter* conquered the New World in two waves, via the South-East Asian route. His theory about South American origins has not yet been proven, but who knows what bacteria still remain to be discovered before Heyerdahl's theory is confirmed by a microbiological voyage of discovery?

Toxoplasmata: fearless cat riders

A 32-year-old woman cuts her wrists with a razor blade that she bought at a discount drugstore. It's the thrill that makes her do it.

A fifteen-year-old racing-car fan crashes into a tree at full pelt. And dies.

A rat drapes itself over the cat's food dish in the kitchen, presenting itself as a delicious meal.

What do these three have in common?

They are all failing to heed the internal signals that aim to preserve the huge community of cells that makes up living creatures. The cells only want the best for us. These three individuals seem to be pursuing interests at odds with those of their bodies — interests that may well have come out of a cat's gut.

Cats' guts are the home of *Toxoplasma gondii*. These tiny little organisms consist of only one cell, and are classed as protozoans — from the Greek meaning 'primitive animals'. They carry much more complex genetic information than bacteria do. Their cell membranes are also constructed differently, and they probably lead more exciting lives.

Toxoplasmata reproduce in the guts of cats. The cat acts as their 'definitive host', while all other animals that toxoplasmata use temporarily as taxis to take them from cat to cat are defined as 'intermediate hosts'. A cat can only get toxoplasmata once in its life, and is a danger to us only during that time of infection. Most older cats already have their toxoplasmata infection behind them, and so cannot harm us anymore. During a fresh infection, toxoplasmata are found in the animal's faeces, and mature in the cat litter for about two days before they are ready to infect their next host. If no cat happens to pass by, but a dutiful, cat-owning mammal comes and cleans the litter, these tiny protozoa seize the opportunity. The microscopic creatures in the cat's faeces can then wait up to five years to infect another 'definitive host', and their intermediate host does not necessarily have to be a human cat-lover. Cats and other animals roam through gardens and vegetable patches, and sometimes get killed. One of the main vectors of toxoplasma infection is raw food. The probability, in percentage terms, of you having toxoplasmata is about as high as your age in years. On a global scale, about a third of humans harbour them.

Toxoplasma gondii are counted as parasites because they cannot live on just any little patch of earth, and absorb water and plant tissue — they need a little patch of organism to live on. As humans, we call such creatures 'parasites', because we get nothing in return from them. At least, nothing positive, like monthly rent or affection. Quite the opposite, in fact: some can harm us by means of a kind of 'human pollution'.

They do not have any overly negative effects on healthy adult hosts. Some people have mild, flu-like symptoms, but most notice nothing. After the acute infection phase, the toxoplasmata move into tiny apartments in our tissue and enter a kind of hibernation state. They will never leave us for the rest of our lives, but they are quiet little lodgers. Once we have been through this, we can never be re-infected. We are already occupied, so to speak.

However, an infection like this can have drastic consequences for pregnant women. The parasites can reach an unborn child via the mother's bloodstream. Its immune system is not yet familiar with them, and is not fast enough to catch them. This does not necessarily always happen, but when it does, it can cause serious damage and even a miscarriage. If the infection is detected early enough, it can be treated with medication. But the chances of that are slim, since most people do not notice when they become infected. And in Germany, for example, a toxoplasma scan is not part of the standard set of pregnancy examinations. If your gynaecologist starts asking strange questions like, 'Do you own a cat?' at your initial pregnancy examination, don't brush it off as meaningless small talk — she's clearly an expert in her field.

Toxoplasmata are the reason your cat's litter should be changed every day if there is a pregnant woman in the house (but not by her!), why raw food should be avoided by mothers-to-be, and why fruit and vegetables should always be washed. Toxoplasmata cannot be transferred from person to person. Infection can only come from the little residents of a freshly infected cat's gut. But, as mentioned earlier, they can survive for a long time, even on the hands of cat owners. Once again, good old hand-washing is the best defence.

So far, so good. All in all, toxoplasmata seem to be unpleasant but otherwise unimportant little critters, if you don't happen to be pregnant. And, for many years, no one paid them much attention — but Joanne Webster's fearless rats changed all that. In the 1990s, when Joanne Webster was a researcher at Oxford University, she devised a simple but ingenious experiment. She placed four boxes in a small enclosure. In one corner of each of these boxes, she placed a small bowl containing a different liquid: rat urine, water, rabbit urine, and cat urine. Even rats that have never seen a cat in their lives avoid cat urine. They are biologically programmed to think, 'If someone peed there who might want to eat you, don't go there.' Furthermore, rodents have a general motto that goes something like this: 'If someone places you in an enclosure with boxes containing urine, be on your guard.' Under normal circumstances, all rats behave the same way. They briefly explore the unfamiliar environment, and then withdraw into one of the boxes with the less-threatening urine in it.

But Webster found there were exceptions: rats that suddenly displayed completely atypical behaviour. They inquisitively explored the whole enclosure, apparently oblivious to risk, even defying their instincts and entering the box containing the bowl of cat urine, and hanging out there for a while. After observing them for longer, Webster was even able to conclude that they seemed to prefer the cat-urine box to the others. Nothing seemed to interest them more than cat pee.

A smell that should have registered in their brain as sign of mortal danger was suddenly perceived as attractive and interesting. These animals had become uninhibited seekers after their own downfall. Webster knew that there was only one difference between these rats and normal specimens — they were infected with toxoplasmata. This was an incredibly clever move on the part of the tiny parasites. They had caused the rats to offer themselves as food to their definitive hosts!

This experiment caused such consternation in the scientific community that it was repeated in other laboratories around the world. Scientists wanted to make sure the results were not a fluke by testing whether their own lab rats would act in the same way if they were infected with toxoplasmata. And they did. The experiment is now considered flawless, and scientifically sound. Scientists also discovered that the change in behaviour was related only to the rats' response to cat urine, while dog urine elicited the expected avoidance behaviour.

These results whipped up a storm of debate: how could such tiny parasites influence the behaviour of little mammals so drastically? To die or not to die — this is a huge question that any organism worth its salt should be able to answer, as long as there is no parasite on the decision-making committee. Surely?

It was not much of a leap from the little mammal to a larger mammal (= human beings). Might it be possible to find human candidates who had succumbed to a kind of 'feed myself to the cat' instinct, in the form of inappropriate reflexes, reactions, and fearlessness? One approach to finding an answer to this question was to test the blood of people who had been involved in traffic accidents. The quest was to find out whether more of the unfortunate road users would turn out to be toxoplasma carriers than members of society who had not been involved in an accident.

The answer was yes. The risk of being involved in a traffic accident is higher among toxoplasma carriers, especially when the infection is in the active early stage rather than the later, dormant stage. Three small initial studies were followed by a large-scale investigation, and all of them confirmed these results. The large-scale study involved taking blood samples from 3,890 Czech army recruits and testing them for toxoplasmata. The recruits were monitored in the following years, and the number of accidents they were involved in was recorded and analysed. Severe toxoplasma infection, in conjunction with a particular blood group (rhesus negative) turned out to be the highest risk factor. Blood groups can indeed play an important role in parasite infections — some groups offer greater protection than others.

But how does our lady with the razor blade fit into all this? Why is she not horrified by the sight of her own blood? Why is the feeling of slicing through her skin, flesh, and nerves not processed as painful, but as thrilling? How can pain have become the hot sauce in the otherwise bland soup of her everyday life?

There are various ways of approaching this question. One of them is to look at toxoplasmata. When we are infected with them, our immune system activates an enzyme (IDO) to protect us from these parasites. It breaks down a substance that the invaders like to eat, forcing them to enter the less active, dormant state. Unfortunately, this substance is also one of the ingredients needed to produce *serotonin*. (Remember: a lack of *serotonin* is associated with depression and anxiety disorders.)

If the brain lacks *serotonin* because IDO has snatched it all away from under the parasite's nose, our mood can be affected negatively. In addition, nibbled-on precursors of *serotonin* can dock onto certain receptors in the brain, and cause symptoms like lethargy, for example. These are the same receptors as those targeted by painkillers — and the result is indifference and sedation. It can take quite drastic measures to drag the brain out of the state of torpor, and to feel emotions keenly again.

Our body is clever. It carries out a risk/benefit analysis: when a parasite needs to be combatted in the brain, the brain's owner is likely to be in a bad mood. Activating IDO is usually a kind of compromise. The body occasionally uses this enzyme to snatch food away from its own cells. IDO is more highly activated during pregnancy — but only near the interface between mother and child. There, it snatches the food away from immune cells. That weakens them — making them react more mildly to the semi-alien presence of the baby.

Would the lethargy triggered by IDO be enough to drive someone to suicide? Or, to put the question another way: what does it take to make people think about killing themselves? Where would a parasite need to start, if it wanted to switch off our natural fear of harming ourselves?

Fear is associated with a part of the brain called the amygdala. Certain fibres run directly from the eyes to the amygdala, and so the mere sight of a spider can trigger an immediate reaction of fear. This connection exists even in blind people whose visual cortex has been damaged by an injury to the back of the head. They no longer 'see' the spider, but still 'feel' it emotionally. So our amygdala plays a major role in the development of fear. If the amygdala gets damaged, a person may become fearless.

Examinations of intermediate toxoplasmata hosts show that the apartments they occupy to hibernate in are mainly found in the muscles and the brain. Those in the brain are found in three locations. In descending order of frequency, they are the amygdala, the olfactory centre, and the area of the brain directly behind the forehead. As we know, the amygdala is responsible for the perception of fear. The olfactory centre could be responsible for the rat's new-found love of cat urine. The third area is slightly more complex.

This part of the brain creates possibilities by the second. If a research subject is wired up to a brain scanner and confronted with question about faith, personality, or morality, or if he is asked to complete complex and challenging cognitive tasks, lively activity is recorded in this region. One theory proposed by brain researchers is that such activity indicates that this area of the brain is drawing up many designs every second. 'I could believe in the religion followed by my parents. I could start licking the desk in front of me during this conference. I could read a book and have a cup of tea. I could dress the dog up in a funny costume. I could film myself singing a jolly song. I could drive my car at 150 kilometres an hour. I could reach for that razor blade ...' Hundreds of possibilities every second — but which will win though, which will be carried out?

So, if you are a parasite with a plan, it makes sense to settle in this part of the brain. From here, it might be possible to promote self-destructive tendencies — weakening the mechanisms that suppress these courses of action.

Researchers wouldn't be researchers if they hadn't come up with the idea of repeating Joanne Webster's experiment with human beings. So they had humans sniff different animals' urine. Men and women who were toxoplasma carriers had a different reaction to the smell of cat wee than those who were parasite-free. Men liked the smell considerably more; women, less.

Smell is one of our most basic senses. Unlike taste, hearing, or vision, smells are not checked out before they make their way to our consciousness. Strangely, we can dream all sensory experiences except smells. Our dreams are always odourless. Truffle pigs know just as well as toxoplasmata that smells can trigger an emotional response. As it happens, truffles smell pretty similar to a sexy truffle-pig male, and when they smell him hiding under the earth, infatuated female truffle pigs will dig and dig until … they discover a disappointingly unsexy fungus for their owner. I think the astronomical price of truffles is more than justified when you consider how frustrating such a find must be for a poor sow. Anyway, the point is that smells can stimulate attraction.

Some shops exploit this phenomenon — economists call it 'scent marketing'. One American clothing manufacturer even uses sex pheromones. You can often see long queues of teenagers outside the store's darkened and pheromone-scented entrance in Frankfurt. If the shopping precinct were closer to areas with free-ranging pigs, some pretty entertaining scenes might result.

So, if another organism can make us perceive smells differently, could it also influence other sensory impressions?

There is a well-known illness whose main symptom is false sensory perceptions: schizophrenia. For example, schizophrenics might feel like armies of ants are crawling all over their backs, although there are no such insects anywhere nearby. They hear voices and obey their commands, and they can be extremely lethargic. About 0.5 to 1 per cent of people are schizophrenic.

There is much about the clinical picture of schizophrenia that is still unclear. Most drugs used to treat it do so by deactivating a signal transmitter in the brain that is overabundant in schizophrenics: dopamine. Toxoplasmata possess genes that influence the production of dopamine in the brain. Not all people who suffer from schizophrenia are parasite carriers — so that can be ruled out as the sole cause — but the proportion of carriers among schizophrenics is about twice that among non-schizophrenics.

Toxoplasma gondii could, therefore, influence us via the fear, smell, and behavioural centres of the brain. A higher risk of being involved in an accident, suicide, or having schizophrenia point towards the fact that an infection does not leave all of us unaffected. It will take some time before discoveries like these have consequences for standard medical practice. Suspicions need to be scientifically proven, and further research into possible treatments is needed. This insistence of science on time-consuming validation processes can cost lives. Antibiotics, for example, did not appear in our pharmacies until decades after they were discovered. But this caution can also save lives. Thalidomide and asbestos could easily have been tested for a little bit longer before entering the market.

Toxoplasmata have the ability to influence us far more than we would ever have thought possible just a few years ago. And they have rung in a new scientific age: an age in which a crude lump of cat faeces can show us how susceptible our lives are to change; an era in which we are just beginning to understand just how complex the connections are between us, our food, our pets, and the microscopic world in, on, and around us.

Is this spooky? Well, maybe a little bit. But isn't it also exciting to see how we are gradually decoding processes that we used to believe were part of our inescapable destiny? This could help us to grab the risks by the scuff of the neck and defy them. Sometimes, it takes nothing more than a shovel for the cat litter, a well-cooked chicken, and properly washed fruit and vegetables.

Worms

There are some little white worms that like to live in our gut. Over many centuries, they have adapted their behaviour to life with us.

Half the world's population has had a visit from these worms at some time or other. Some people never even notice. For others, it's an embarrassing infestation they'd rather not talk about. If you look at just the right moment, you can even see them giving us a wave out of the anus. They are a centimetre to a centimetre-and-a-half long, are white, and sometimes have a pointed end. They look a little like the vapour trails left by jets in the sky, with the exception that they don't get longer and longer. Anyone who has a mouth and a finger can get these parasites, known as threadworms or sometimes pinworms. Finally, there's an advantage to being fingerless and/or mouthless!

Let's put the cart before the horse, or rather, the worm. A 'pregnant' lady-worm, looking for a place to lay her eggs, wants to make sure they will have a secure future. And such a place is not so easy to reach. A worm egg has to be swallowed by a human being, then hatch in the small intestine, so that it can reach the large intestine by the time it is a grown-up worm. But our mother-worm-to-be lives in the lower reaches of the digestive tract, with everything moving in the wrong direction for her needs. So she wonders how she is ever going to get back to her host's mouth to give her eggs the start in life that they need. Here, she makes use of the only kind of intelligence such a creature has at her disposal — the intelligence of adaptation. Whether this is the origin of the word 'arse-crawler' or not is open to question.

Female threadworms know when we are still, lying down, and too comfortable to rouse. That's exactly when they set off towards the anus. They lay their eggs in the many little creases around the anus, and wriggle around until it starts to itch. They then slip quickly back inside the gut, because experience has taught them: soon a hand will appear and finish off the job. Under the bedclothes, the hand heads for the backside, targeting that itch. The same neural pathways that passed on the itch now give the instruction to scratch. We obey the instruction, thereby providing the threadworm's children with an express connection to the mouth area.

When are we least likely to go and wash our hands after scratching our bottom? When we are oblivious to all this action because we are asleep, or too sleepy to get up and head for the bathroom. And that is threadworm egg-laying time. It's clear what that next dream about sticking your finger in a yummy chocolate cake will mean — those eggs are heading to their ancestral home: your mouth. That might sound gross, but it's not so very different from eating chickens' eggs. (Only they are bigger, and usually cooked.)

Organisms that move into our gut uninvited, and implement their family planning there, get a bad rap from us. And we often avoid talking about them with others. It's as if we feel we have been a bad manager for our body, failing to lay down the law properly, and letting any old strangers take up residence without interviewing them first. But threadworms are not just any old strangers — they are guests who wake the manager in time for morning exercises, and then give their host a massage to stimulate the immune system. Furthermore, they steal very little of our food.

It is not good to keep them as permanent guests — but once in a lifetime is fine. Scientists suspect that when kids have had worms, they are less likely to contract severe asthma and diabetes in later life. So, welcome, Mr and Mrs Threadworm, come on in! But don't outstay your welcome, please. An uncontrolled attack of worms can have three rather unwelcome consequences:

1. Lack of a good night's sleep can lead to concentration problems, nervousness, or irritability during the day.
2. What the worms don't want is to lose their way — and nor do we. When worms get into places they don't belong, they have to go. Who wants a threadworm with a bad sense of direction, after all?

3. Sensitive guts, or those containing overactive worms, can become irritated. Worms have a tendency to cause irritation anyway. There are many problems this can cause: not going to the toilet often enough, going to the toilet too often, abdominal cramps, headaches, nausea, or none of the above.

If a worm host has any of these symptoms, a visit to the doctor is essential. The doctor will ask you to put sticky tape to a use you never learned in arts and crafts lessons at primary school. Some doctors are more charming about it than others, but, in essence, what they will tell you to do is: spread your cheeks, stick the tape to your anus and the surrounding area, pull it off again, bring it to the surgery, and hand it over to the receptionist.

Worm eggs are small and round, and adhere well to sticky tape. Searching for eggs on Easter Sunday morning would be a lot more efficient if you had a great big egg-magnet that attracted all the eggs from the garden. Since worm eggs are so much smaller than Easter eggs, it makes sense to use a trick like this. The sticky-tape egg hunt has to take place in the morning, as that's when most eggs are there. And it is not a good idea to flush out or sweep clean the worm garden before you hunt for the eggs. So the first thing that comes into contact with the region in the morning should be those strips of sticky tape.

The doctor will examine the fruits of your labours under the microscope, hunting for little oval eggs. If they are already developing into larvae, they will have a line down the middle. The doctor can then prescribe the right medication, and your pharmacist will help you win the battle to get rid of your unwanted guests. The typical medication prescribed — let's call it Mebendazole, for the sake of argument — works on the tit-for-tat principle we all know from kindergarten: if you bother my gut, I'll bother yours.

The medication makes its way from mouth to rectum, and

meets our renegade squatters along the way. Mebendazole is much more harmful to a worm's gut than it is to ours. It places the worms on a forced diet, denying them all access to sugar. Sugar is the stuff of life for worms, so this will be the last diet they ever go on. It's a bit like trying to get rid of unwanted guests by not offering them anything more to eat.

Threadworm eggs live for a long time. If you have worms and can't keep your hands away from your mouth, you should at least try to reduce the number of eggs in the area to a minimum. Bedclothes and underwear should be changed every day, and washed at 60°C or hotter; regular hand washing is essential; and intense itching might be better treated with creams than by scratching. My mother is convinced that worms can be eliminated by swallowing a whole clove of garlic once a day. I have not been able to find any scientific studies to prove this, but neither are there any studies about what temperatures necessitates the wearing of warm jackets, and my mother is always right about that! If all this fails, do not despair. Go back to your doctor, and be proud of having such an inviting gut.

Of Cleanliness and Good Bacteria

We always want to protect ourselves from harm. Few people would choose to have salmonella or a nasty *Helicobacter*. Even though we have not yet identified them all, we know already that we'd rather not have chubby bacteria, or microbes that cause diabetes or depression. Our greatest protection against them is cleanliness. We are careful about eating raw food, kissing strangers, and washing our hands to rid them of anything that might spread disease. But cleanliness is not always what we imagine it to be.

Cleanliness in our gut is something akin to cleanliness in a forest. Even the most conscientious of cleaners would not dream of taking a mop to the forest floor. A wood is 'clean' if the beneficial plants it contains are in healthy equilibrium. We can help the forest along — by sowing seeds and hoping new plants will take root. We can identify favourite or useful plants in the forest, and nurture them to help them grow and multiply. Sometimes, there are nasty pests. Then, careful consideration is in order. If the situation is desperate, chemicals might be the answer. As their name implies, pesticides are great at killing pests, but it is not a great idea to spray them round like air freshener.

Clever cleanliness begins with our everyday routines — but what is well-advised caution, and what is excessive hygiene? There are three main tools for keeping our insides clean. Antibiotics can rid us of acute pathogens, while prebiotic and probiotic products promote beneficial elements. '*Pro bios*' means 'for life'. Probiotics are edible living bacteria that can make us healthier. '*Pre bios*' means 'before life' — these are foodstuffs that pass undigested into the large intestine, where they feed our beneficial bacteria so that they thrive better than bad bacteria. '*Anti bios*' means 'against life'. Antibiotics kill bacteria, and are our saviours when we have picked up a pack of bad bacteria.

Everyday cleanliness

The fascinating thing about cleanliness is that it is mostly in the head. A peppermint tastes fresh; clean windows look clear; and there's nothing better than the lovely feeling of slipping into a freshly made bed after a hot shower. We like the smell of clean things. We like to run our hands over smooth, polished surfaces. We find comfort in the idea that we are protected from an invisible world of germs if we use enough disinfectant.

In Europe, 130 years ago, it was discovered that tuberculosis is caused by bacteria. This was the first time the public took notice of bacteria — and they were seen as bad, dangerous, and, most worryingly, invisible. It was not long before new regulations were introduced in European countries: patients were isolated so they could not spread their germs; spitting was forbidden in schools; close physical contact was discouraged; and warnings were issued against 'the communism of the towel'! People were even advised to limit kissing to 'the erotically unavoidable'. This might sound funny to us today, but those ideas put down deep roots that can still be felt in modern Western society: spitting is still frowned upon; we are still reluctant to share towels and toothbrushes; and we keep a greater physical distance in our dealings with others than do most cultures in the non-Western world.

Preventing deadly disease by banning pupils from spitting at school seemed like a simple and effective idea. Consequently, we internalised it in our culture as a social rule. Those who did not comply were despised as a danger to everyone's health. This attitude was passed on from parent to child, and public spitting became a social taboo. Cleanliness really was thought to be next (in importance) to godliness; people craved a sense of order in a life full of chaos. The anthropologist Mary Douglas summed this up in her book *Purity and Danger* with the phrase 'dirt is matter out of place'.

Bathing as a way of keeping the body clean was a privilege of the rich even up to the beginning of the twentieth century. But it was around this time that dermatologists in Germany began to call for 'a bath a week for every German!' Large companies built bathhouses for their employees, and encouraged personal hygiene by issuing them with free towels and soap. The tradition of the weekly bath did not really take hold until the 1950s. Then, typical families took their bath on a Saturday evening, one after another in the same bathwater, and hard-working Dad often got to go in the tub first. Originally, personal cleanliness meant ridding the body of unpleasant smells and visible dirt. As time went on, this concept became increasingly abstract. It's hard for us today to imagine this once-a-week family bathing-routine. We spend money on disinfectants to get rid of things we can't even see. The surface in question looks exactly the same after cleaning as it did before — yet just knowing it is clean is extremely important to us.

The news media regale us with horror stories about dangerous flu viruses, multidrug-resistant superbugs, and EHEC-contaminated food. When food-contamination scares are in the news, like the EHEC outbreak in Germany in 2011, some people react by giving up cucumbers; others type 'full body decontamination shower' into Google. Different people deal with fear in different ways. Dismissing this as hysteria is too easy — it makes more sense to try to understand where these fears come from.

Fear-driven hygiene involves either trying to clean everything away, or kill it off. We don't know what it might be, but we assume the worst. When we clean obsessively, we do indeed get rid of everything — both bad and good. This cannot be a good kind of cleanliness. As it happens, the higher the hygiene standards in a country, the higher that nation's incidence of allergies and autoimmune diseases. The more sterile a household is, the more its members will suffer from allergies and autoimmune diseases. Thirty years ago, about one person in ten had an allergy; today, that figure is one in three. At the same time, the number of infections has not fallen significantly. This is not smart hygiene. Research into nature's huge range of bacteria has led to a new understanding of what cleanliness should mean. We have now moved on from the time when it was defined as the attempt to kill off potential dangers.

More than 95 per cent of the world's bacteria are harmless to humans. Many are extremely beneficial. This means that disinfectants have no place in a normal household — only if a family member is sick, or if the dog poops on the carpet. In the case of the latter, there are no holds barred: feel free to use steam cleaners, disinfectant by the bucket-load, and a small flame-thrower, perhaps. That might even be fun. But if the floor is covered just in dirty footprints, all you need is water and a drop of cleaning fluid. That combination is already enough to reduce the bacteria population of your floor by 90 per cent. And that gives the normal, healthy population of the floor a chance to recolonise the territory. What remains of any harmful elements is so little as to be negligible.

The aim of cleaning, then, should be to reduce bacteria numbers, but not to eliminate them. Even harmful bacteria can be good for us when the immune system uses them for training — a couple of thousand salmonella bacteria in the kitchen sink provide our immune system with the opportunity to do a little sightseeing. Salmonella become dangerous only when they turn up in greater numbers. In general, bacteria get out of hand only when they encounter perfect conditions: a protected location that is warm and moist, and a supply of delicious food. There are four reliable strategies for keeping them in check: dilution, temperature change, drying, and cleaning.

Dilution

Dilution is a technique we also use in the laboratory. We dilute bacteria with fluids, and we administer drops with different concentrations of bacteria to wax moth larvae, for example. Wax moth larvae change colour when they get sick. This makes them a good indicator of the concentration of bacteria required to cause illness. For some, it's a little as 1,000 per drop of fluid; for others, as many as 10 million.

One example of bacteria dilution in the home is washing fruit and vegetables. This dilutes most soil-dwelling bacteria to such a low concentration that they become harmless to humans. Koreans add a little vinegar to the water to make it slightly acidic and just that bit more uncomfortable for any bacteria. Airing a room is also a dilution technique.

But if you dilute the bacteria on your plates, cutlery, and cutting board nicely with water, and then wipe them over with a kitchen sponge before putting them away, you may as well have licked them clean with your tongue. Kitchen sponges offer the perfect home for any passing microbe — nice and warm, moist, and full of food. Anyone looking at a kitchen sponge under the microscope for the first time usually wants to curl up on the floor in the foetal position, rocking back and forth in disgust.

Kitchen sponges should only be used for cleaning the worst of the dirt off — plates, cutlery, and so on should then be rinsed briefly under running water. The same is true for tea towels or drying-up cloths if they never get a chance to dry out. They are more useful for spreading a nice, even layer of bacteria on your utensils than for drying them. Sponges and cloths should be thoroughly wrung out and allowed to dry — otherwise they become the perfect place for moisture-loving microbes.

Drying

Bacteria cannot breed on dry surfaces. Some cannot survive there at all. A freshly mopped floor is at its cleanest after it has dried. Armpits that are kept dry by antiperspirants are less cosy homes for bacteria — and fewer bacteria produce less body odour. Drying is a great thing. If we dry food it keeps for longer before it rots. We use this to our advantage: just think of foodstuffs like pasta, muesli, crispbread, dried fruit (such as raisins), beans, lentils, and dried meats.

Temperature

The environment is refrigerated naturally once a year. From a bacteriological point of view, winter is the real spring clean. Refrigerating food is an extremely important part of our daily lives. But a fridge contains so much food that it remains a paradise for bacteria even at low temperatures. The optimum temperature for your fridge is something below 5°C.

Moving on to another household appliance: most washing-machine programmes use the dilution principle to clean our clothes, and that is sufficient. However, damp kitchen cloths, a load of underpants, or sick people's laundry should be washed at 60°C or more. Most *E. coli* bacteria are killed by temperatures above 40°C; 70°C is enough to kill off tougher salmonella bacteria.

Cleaning

'Cleaning' means removing a film of fats and proteins from surfaces. Any bacteria living in it or under it will be removed, along with the film. We usually use water and cleaning fluid to achieve

Fig.: *Bacteria trapped in iodine crystals*

this. Cleaning is the technique of choice for all living spaces, kitchens, and bathrooms.

This technique can be taken to the absolute extreme. That is sensible when you are manufacturing medical drugs which are destined to be pumped straight into a patient's veins — such as infusions — that need to be absolutely free of all bacteria. Pharmacological laboratories achieve this by using iodine, for example. Iodine can be made to sublimate — that is, to transition from a solid, crystalline state to a gaseous vapour in the presence of heat, without passing through a liquid phase. So pharmacologists heat up iodine until the entire production lab is veiled in a blue vapour.

It sounds like the simple principle of the steam cleaner, but there is more to it than that. Iodine can also desublimate. To make this happen, the room is cooled, and the vapour immediately recrystallises. Millions of tiny crystals form on all surfaces and even in mid-air, trapping every microbe as they do so, locking them up in a crystal prison as they fall to the ground. Workers pass through several airlocks and disinfection chambers, dress up in germ-free bodysuits, and sweep up the iodine crystals.

In principle, we use the same system when we use hand cream — we trap microbes in a film of fat, and hold them captive there. When we wash the film off, we rinse the bacteria away with it. Since our skin produces a natural coating of fat, it's often enough to achieve this effect by using soap and water. Some of the fat layer remains, aiding its replenishment after washing. But too-frequent hand washing makes no sense — and the same is true of too-frequent showering. If the protective fat layer is rinsed away too often, our unprotected skin is exposed to the environment. This gives odour-producing bacteria a better foothold, making us smell more pungently than before: a vicious circle.

New methods

A team from Ghent in Belgium is currently trying out a brand-new method: the researchers are attempting to use bacteria to combat body odour. They disinfect volunteers' armpits, spread them with odourless bacteria, and start the stopwatch. After a couple of minutes, the subjects are allowed to put their shirts back on and go home. The volunteers return repeatedly to the laboratory, where experts sniff their armpits. The initial results are quite promising — the odourless bacteria manage to banish their smelly colleagues in many of the subjects' armpits.

Not far away, across the border in Germany, the same method is in use in the public toilets of the small town of Düren. A company is using a mixture of bacteria to clean the toilets. The odourless bacteria occupy the places normally colonised by the bugs that create that all-too-familiar public-toilet smell. The idea of using bacteria to clean public conveniences is a brilliant one. Unfortunately, the company refuses to reveal the recipe for its cocktail of bacterial cleaners, so a scientific evaluation is impossible. However, the town of Düren seems to be faring very well with this experiment.

These new ideas about the use of bacteria illustrate one thing very clearly: cleaning does not mean annihilating all bacteria. Cleanliness is a healthy balance of sufficient good bacteria and a few bad ones. That means: smart protection against real dangers, and sometimes deliberate contamination with good bugs. With this in mind, we can perhaps better appreciate the wisdom of observations like that of the American writer, Suellen Hoy, who says, 'From the perspective of a middle-class American woman (also a seasoned traveller) who has weighed the evidence, it is certainly better to be clean than dirty.'

Antibiotics

Antibiotics are reliable killers of dangerous pathogens. And their families. And their friends. And their acquaintances. And distant acquaintances of their acquaintances. That makes them the best weapon against dangerous bacteria — and the most dangerous weapon against good bacteria. But who is it who manufactures most antibiotics? It's bacteria. Huh?

Antibiotics are the weapons used by both sides in the war between fungi and bacteria.

Ever since researchers discovered all this, pharmaceutical companies have been intensively farming bacteria. Huge tanks (with a capacity of up to 100,000 litres) are used to grow so many bacteria that their numbers can hardly be expressed in any meaningful figures. They produce antibiotics, which are purified and pressed into little tablets. The product sells well, especially in the United States. When researchers were planning a study of the effect of antibiotics on the flora of the gut, they were able to find only two people in the entire San Francisco Bay Area who had not taken antibiotics in the previous two years. In Germany, one person in every four takes antibiotics once a year on average. The main reason for taking them is 'colds'. This is like a knife in the heart of any microbiologist. Colds are often not even caused by bacteria, but by viruses! Antibiotics work in three different ways: by filling bacteria with holes, by poisoning bacteria, or by destroying bacteria's ability to reproduce. They have no effect on viruses at all.

So taking antibiotics to cure a cold is usually a complete waste of time. If they do bring about an improvement, this is due either to the placebo effect or to the work of the immune system in combatting the cold virus. The senseless use of antibiotics does, however, kill many helpful bacteria, which can be harmful in itself. To avoid this, doctors can perform a *procalcitonin* test, which indicates whether a cold is caused by a bacterial or viral infection. It should be noted that, in Germany at least, such a test is not covered by most health insurers. But it should still be considered, especially in the case of children.

There is no reason not to take antibiotics when it is medically appropriate to do so. The benefits certainly outweigh the disadvantages — for example, in cases of severe pneumonia, or for helping children get over a particularly bad infection with no long-term damage. In such cases, those little tablets can save lives. Antibiotics stop bacteria from reproducing, the immune system kills off any remaining pathogens, and we soon start to feel better. We have to pay a price for this, but, in the final reckoning, it's a good deal.

The most common side effect is diarrhoea. Even those who don't get diarrhoea may notice that they leave a rather larger deposit in the morning toilet bowl than usual. To tell it like it is: that is a large portion of dead gut bacteria. The antibiotic tablet does not travel directly from mouth to blocked-up nose. It descends into the stomach and then on to the gut. Before it graduates from there into the bloodstream — and then to the nose, among other places — it peppers our microbe collection with holes, poisons our gut bacteria, and makes them sterile. The result is a formidable battlefield, which you can see the next time you go to the toilet.

Antibiotics can alter our gut flora significantly. Our microbe collection becomes much less diverse, and the abilities of the bacteria in it can change — for example, the amount of cholesterol they can absorb, their ability to produce vitamins (like skin-friendly vitamin H), and the type of foodstuffs they can help us digest. Preliminary studies carried out at Harvard and in New York have shown that the two antibiotics metronidazole and gentamycin cause particularly hefty changes in the flora of the gut.

Antibiotics can be problematic for children and old people. Their gut flora are already less stable and less able to recover after treatment with antibiotics. Research in Sweden has shown that the gut bacteria of children were still significantly altered two months after taking antibiotics. Their guts contained more potentially harmful bacteria, and fewer beneficial types, such as *Bifidobacteria* and Lactobacilli. The antibiotics used were ampicillin and gentamycin. The study involved only nine children, which means it is not particularly meaningful in scientific terms — but it remains the only study of its kind so far. So the results should be accepted with a degree of caution.

A more recent study of pensioners in Ireland revealed a clear dichotomy. Some gut landscapes recovered very well after a course of antibiotics; others remained permanently altered. The reasons for this are still completely unclear. The ability to return quickly to a stable state following an extreme experience is described with the same word by gut researchers and psychologists: resilience.

Studies of the long-term effects can still be counted almost on one hand — although antibiotics have been in use for more than fifty years. The reason for this is technological: the equipment necessary for such investigations has only been around for a couple of years. The only long-term effect that has been scientifically proven is drug resistance. As long as two years after taking antibiotics, bad bacteria are still present in the gut, telling their great-great-great-...-grandchildren stories about the war.

These are the ones that resisted the antibiotics and survived. And with good reason. They developed resistance strategies — for example, by installing tiny pumps in their cell walls, to pump the antibiotic out like emergency workers pumping water out of a flooded cellar. Some bacteria prefer to disguise themselves, so that the antibiotics cannot recognise their surface and cover them with holes. Yet others use their ability to split things — they build tools to cut the antibiotics to pieces.

The thing is that antibiotics rarely kill all bacteria. They kill certain communities of them — depending on the toxins they use.

But there will always be some bacteria that survive and become experienced fighters. In the case of serious illness, these fighters can become a problem. The more resistances they have developed, the more difficult it is to get them under control again with antibiotics.

Every year, many thousands of people die in the West because they are infected with bacteria that have developed resistances that no drug can counter. When their immune systems are compromised — for example after an operation, or if the resistant bacteria have got out of hand after a long course of treatment with antibiotics — patients can find themselves in real danger. Very few new antibiotic drugs are in development, for the simple reason that it is not very profitable for pharmaceutical companies to invest in them.

So here are five pieces of advice for anyone who wants to keep out of unnecessary antibiotic gut wars:

1. Do not take antibiotics unless it is really necessary to do so. And if you do take them, you should not take them for too long, or too short a time. Long enough for you to really shake off the bacteria that are making you ill and kill off any lingering colonies that could cause the infection to flare up again. But short enough not to kill off so many bacteria that you create space for the very strains you don't want to give a home to, namely those that are resistant to the antibiotic you are taking. There is currently a debate among scientists about whether it might be sensible to stop a course of antibiotics as soon as the patient subjectively begins to feel better.
2. Buy organically farmed meat. Drug resistances differ from country to country, and it is shocking to see how often they correspond to the antibiotics used in large-scale animal farming. In countries like India or Spain, for example, there is almost no regulation of the amount of antibiotics given to animals. This turns the animal's guts into giant breeding zoos for resistant bacteria, and people in such regions have significantly more infections with multidrug-resistant strains. In Germany, regulations do exist, but even there the rules are ridiculously vague. This allows many vets to make a lot of money in the semi-legal 'antibiotics trade'.
3. It was not until 2006 that the EU banned the use of antibiotics in animal feed as 'performance enhancers'. The kind of 'performance' being 'enhanced' here is the ability of an animal not to die from infections in a crowded and dirty pen. Antibiotics are a great way to 'enhance' the 'performance' of living. Animals farmed organically are allowed to receive only specified amounts of antibiotics. If those amounts are exceeded, the animal is sold for 'normal' meat, without the

organic label. If possible, it is worth spending that little extra — to prevent resistance-breeding zoos and for your peace of mind … and 'peace of gut'. The dividends are not immediate, but it is an investment in a better future for us all.

4. Wash fruit and vegetables thoroughly. This is also connected with animal farming. Animal faeces are a popular fertiliser. Liquid manure is used in vegetable fields. In Germany, fruit and vegetables are not routinely tested for residues of antibiotics — and certainly not for multidrug-resistant gut bacteria. Milk, eggs, and meat are, however, tested to make sure they don't exceed certain limits. So err on the side of caution, and wash your fruit and veg one extra time if you are not sure. Even tiny amounts of antibiotics can help bacteria develop resistances.

5. Take care abroad. One holidaymaker in four returns home carrying highly resistant bacteria. Most of these bugs disappear in a few months, but some lurk around for much longer. Special care should be taken in bacterial-problem countries like India. In Asia and the Middle East, you should wash your hands regularly, and clean fruit and vegetables thoroughly — if necessary, with boiled water. Southern Europe also has its problems. 'Cook it, peel it, or leave it' is not only a good rule for avoiding diarrhoea; it also protects against unwanted resistant souvenirs for you and your family.

Are there alternatives to antibiotics?

Plants (fungi, such as the penicillin fungus, are not plants but *opisthokonts*, like animals) produce antibiotics that have functioned for centuries without causing resistance. When parts of plants snap off or become perforated, they need to produce antimicrobial substances at the location of the damage. If they did not do this, they would immediately become a feast for any bacteria in the vicinity. Pharmacies sell such concentrated plant antibiotics to treat developing cold symptoms, urinary infections, and inflammations in the mouth and throat. Some products contain mustard seed or radish-seed oil, for example, or chamomile and sage extract. Some have the ability to reduce the numbers of not only bacteria, but also of viruses. That leaves our immune system with less to contend with, giving it a better chance of dealing with pathogens.

Such plant-based remedies are not the solution for serious illnesses, or for illnesses that drag on with no noticeable improvement. In such cases, they can even be harmful because they encourage us to wait too long before turning to powerful antibiotics. In recent years, the incidence of heart and ear damage in children has increased; this is often due to the behaviour of parents who want to protect their children from too much exposure to antibiotics. Such a decision can have damaging consequences. A well-informed doctor will not prescribe antibiotics for every little thing — but will tell you in no uncertain terms when they are really necessary.

Our relationship with antibiotics is an arms race: we use them to arm ourselves to the hilt when faced with dangerous bacteria, and they respond by arming themselves with even more dangerous resistances. Medical researchers should really be developing the next generation of weapons in this race. Every one of us accepts a trade-off when we take these drugs. We agree to sacrifice our good bacteria, in the hope of getting rid of the bad. In the case of a minor cold, that's not a good deal; for serious illnesses, it's a trade that pays off.

There is no species-protection programme for gut bacteria, so we can be pretty certain that we have annihilated many family heirlooms since the discovery of antibiotics. The places they leave vacant should be colonised by the best candidates possible — probiotics, for example. They help the gut to return to a state of healthy equilibrium after danger has been averted.

Probiotics

We swallow many billions of living bacteria every day. They thrive on raw food, a few survive on cooked food, we nibble on our fingers without even thinking about it, we swallow our own mouth bacteria, or we kiss our way through the bacterial landscapes of others. A small proportion even survive the acid bath of the stomach, and the aggressive digestive process, to reach our large intestine alive.

No one knows about the majority of these bacteria — they presumably do us no harm, or perhaps even benefit us in some way we have not yet discovered. A few are pathogens, but usually cannot harm us because their numbers are just too small. Only a fraction of these bacteria have been thoroughly checked out by scientists and given the official seal of 'good'. These bacteria can proudly call themselves probiotics.

We often read the word 'probiotic' on the yoghurt we find on our supermarket shelves, without having any real idea what it means or how they work. Most of us will recall TV ads telling us they strengthen our immune system, or showing a constipated aunt relieved of her woes and recommending this brand to everyone she knows. This all sounds good. You don't mind spending a little extra on a healthy product like that. And before you know it, those probiotics are in your shopping basket, then in your fridge, and eventually in your mouth.

People have been eating probiotic bacteria since time immemorial. Without them, we would not exist. A group of South Americans had to learn this through bitter experience: they had the clever idea of taking pregnant women to the South Pole to have their babies. The plan was that the babies born there could stake a claim to any future oil reserves, as 'natives' of the region. But the babies did not survive. They died soon after birth, or on the way back to South America. The South Pole is so cold and germ-free that the infants simply did not receive the bacteria they needed to survive. The normal temperatures and bacteria the babies encountered after leaving the Antarctic were enough to kill them.

Helpful bacteria are an important part of our life, and we are constantly surrounded and covered by them. Our ancestors had no idea of their existence, but intuitively did the right thing: protecting their food from the bacteria that made it rot by handing it over to the care of good bacteria. They used bacteria to preserve their food. Every culture in the world includes traditional dishes that rely on the help of microbes for their preparation. Germany has its sauerkraut, pickled gherkins, and sourdough pretzels. The French love their crème fraîche. The Swiss have their 'holey' cheese. Salami and preserved olives come from Italy. Turks swear by a salty yoghurt drink called ayran. None of these delicacies would exist if it weren't for microbes.

There are many, many examples from Asian cuisine: soya sauce, kombucha drinks, miso soup, Korean kimchi, Indian lassi, and African fufu ... the list is endless. All these foods rely on bacteria for a process we call 'fermentation'. The process often results in the production of acid, which makes the yoghurt or vegetables taste sour. This acid, along with the many good bacteria, protects the food from dangerous microbes. Fermentation is the oldest and healthiest way of preserving food.

The bacteria used in this technique were as varied around the world as the foods they helped to produce. The soured milk drunk in rural Germany was made using different bacteria from the ayran enjoyed in Anatolia. In the warmer countries of the south, bacteria were used which prefer to work under higher temperature conditions; in the chilly north, bacteria were chosen which like to do their job at room temperature.

Yoghurt, soured milk, and other fermented products came about by accident. Someone left the milk outside, bacteria found their way into the churn (either directly from the cow or from the air during milking), the milk thickened, and a new kind of food had been invented. If a particularly delicious yoghurt bacteria jumped into the mix, people added a spoonful of that yoghurt to the next batch to make more of the same. Unlike today's yoghurt products, however, traditional types were the work of whole teams of different bacteria — not just selected individual species.

The diversity of bacteria in fermented foods has fallen sharply. Industrialisation has resulted in standardised production processes, using single bacteria species isolated in laboratories. Today, milk is briefly heated soon after it leaves the udder, to kill off any potential pathogens. But this also kills any potential yoghurt bacteria. That's why you can't just leave modern shop-bought milk to go sour in the hope that it will eventually turn into yoghurt.

Many foods that used to be full of bacteria are now preserved using vinegar — most pickled gherkins, for example. Some things are fermented using bacteria, but are then heated to kill off the microbes — shop-bought sauerkraut, for instance. Fresh sauerkraut is usually only sold in specialist healthfood stores these days.

Scientists in the early twentieth century suspected that good bacteria were of great benefit to us. That is when Ilya Metchnikoff appeared on the yoghurt scene. The Nobel Prize winner spent his time observing Bulgarian mountain peasants. He realised they often lived to be a hundred years old or more, and were unusually contented. Metchnikoff suspected the key to their longevity lay in the leather bags they used to transport the milk from their cows. The peasants had to walk long distances, and their milk often turned sour or transformed into yoghurt in the bags before they reached home. Metchnikoff became convinced that the secret to their long lives was their regular consumption of this bacterial product. In his book *The Prolongation of Life*, he asserted the claim that good bacteria can help us live longer, better lives. From then on, bacteria were no longer just anonymous components of yoghurt, but important promoters of health. However, Metchnikoff's timing could hardly have been worse. Shortly before, it had been discovered that bacteria cause disease. Although the microbiologist Stamen Grigorov identified the bacterium described by Metchnikoff as *Lactobacillus bulgaricus* in 1905, he soon turned his efforts to the fight against tuberculosis. The successful use of antibiotics in fighting disease from around 1940 meant that the idea was fixed in most people's minds: the fewer bacteria, the better.

We have babies to thank for the fact that Ilya Metchnikoff's idea and Grigorov's bacillus eventually found their way onto our supermarket shelves. Mothers who were unable to breastfeed their babies found they often had a problem with their bottle-fed offspring. The babies were getting diarrhoea much more often than they should. This took the formula milk industry by surprise, because they had taken care to make sure their product contained the same substances as real breast milk. What could be missing? The answer was, of course, bacteria! Bacteria that live on milky nipples, and which are particularly common in the guts of breastfed babies: *Bifidobacteria* and Lactobacilli. They break down the sugar in milk (lactose) and produce lactic acid (lactate), so they are classified as lactic-acid bacteria. A Japanese scientist used *Lactobacillus casei Shirota* to create a special yoghurt, which mothers could initially buy only in pharmacies. When they fed it to their babies every day, the infants contracted diarrhoea much less frequently. Industrial food research returned to Metchnikoff's idea — with baby bacteria and more modest aims.

Most normal yoghurt contains *Lactobacillus bulgaricus*, although it is not necessarily precisely the same sort as that in the yoghurt of Bulgarian peasants. Today, the species discovered by Stamen Grigorov is more correctly known as *Lactobacillus helveticus spp. bulgaricus*. These bacteria are not particularly good at resisting digestion, and only a small number of them reach the large intestine alive. That is not so important for some effects on the immune system — often, just the sight of a few empty bits of bacteria wall is enough to prompt our immune cells into action.

Probiotic yoghurt contains bacteria that researchers were inspired to use by the case of the bottle-fed babies with diarrhoea. They are meant to reach the large intestine alive. Examples of bacteria that can resist being digested are *Lactobacillus rhamnosus*, *Lactobacillus acidophilus*, or the abovementioned *Lactobacillus casei Shirota*. The theory is that a living bacterium will have a greater effect on the gut. There are studies that show their effects, but they are not sufficient to satisfy the European Food Safety Authority. It has banned companies like Yakult and Actimel from claiming in their advertising that their products promote health.

These doubts are compounded by the fact that it is not always possible to know whether enough probiotic bacteria are reaching the large intestine alive. A break in the cold chain, or a person with a particularly acidic stomach or slow digestion, might kill off these microbes before they reach their intended destination. That is not harmful, of course, but it means that consuming a probiotic yoghurt may have no effect that a normal yoghurt doesn't also have. To make any difference to the huge ecosystem in our gut, about a billion (10^9) bacteria need to make it through the system and arrive there intact.

To summarise: any yoghurt may be good for you, although not everyone can tolerate milk protein or too much animal fat. The good news is that there is a world of probiotics beyond yoghurt. Researchers are busy in their laboratories examining selected bacteria. They dribble bacteria directly onto gut cells in petri dishes, feed mice with microbial cocktails, or get volunteers to swallow capsules full of living micro-organisms. Probiotic research has roughly defined three areas in which our good bacteria can display fascinating abilities.

1. Massaging and pampering

Many probiotic bacteria take good care of our gut. They possess genes that enable them to produce small fatty acids like *butyrate*. This soothes and pampers the villi in the gut. Pampered villi are much more stable and likely to grow bigger than unpampered ones. The bigger the villi grow, the better they are at absorbing nutrients, minerals, and vitamins. The more stable they are, the less rubbish they let through. The result is that our body receives more nutrients and fewer damaging substances.

2. Security service

Good bacteria defend our gut — it is, after all, their home, and they do not willingly surrender their territory to bad bacteria. Sometimes they defend the gut by occupying the very places that pathogens like to infect us most. When a bad bacterium turns up, it finds them sitting in its favourite place, with satisfied grins on their face and their handbags on the next seat, leaving no room for anyone else to take up residence. Should that signal not be explicit enough — no problem! Security-service bacteria have more tricks up their sleeves. For example, they can produce small amounts of antibiotics or other defensive substances that drive unfamiliar bacteria out of their immediate vicinity. Or they use various acids, which not only protect yoghurt or sauerkraut from rotting bacteria, but make our gut a less inviting environment for bad bacteria. Another possibility is to snatch their food away (anyone with siblings may be familiar with this strategy). Some probiotic bacteria seem to have the ability to steal bad bacteria's food right from under their noses. Eventually, the bad guys have had enough, and give up.

3. Good advice and training

And, last but not least, the best experts in all things bacterial are bacteria themselves. When they work together with our gut and its immune cells, they provide us with insider information and useful advice: What do the different bacteria's outer walls look like? How much protective mucous is needed? What quantity of bacterial-defence substances (*defensins*) should the gut cells produce? Does the immune system need to be more active in its reaction to foreign substances, or sit back and accept newcomers?

A healthy gut contains many probiotic bacteria. We benefit every day and every second from their abilities. But, sometimes, our bacterial community faces attack. That can be from antibiotics, a bad diet, illness, stress, and many, many other causes. Our gut is then less well protected and receives less good advice. When that is the case, we can be thankful that some of the results of laboratory research have made it onto pharmacy shelves. Living bacteria are available that can be used like temporary workers brought in to help during times of heavy workloads.

Good for treating diarrhoea. This is the number-one use for probiotics. Gastroenteritis (stomach flu) and diarrhoea caused by taking antibiotics can be helped using various pharmacy-bought bacteria. They can reduce the length of such a bout of diarrhoea by about a day. At the same time, they are almost free of side effects — unlike most diarrhoea medications. That means they are particularly suitable for small children and old people. In conditions like *ulcerative colitis* and irritable bowel syndrome, probiotics can increase the intervals between diarrhoea attacks or inflammatory flare-ups.

Good for the immune system. For people who tend to get sick very often, it can be a good idea to try different probiotics — especially in times when colds are rife. If that is too expensive, eating a pot of yoghurt a day may be enough, since bacteria don't necessary have to be alive to trigger some mild effects. Studies have shown that old people and high-performance athletes, especially, are less prone to catching colds if they take probiotics regularly.

Possible protection against allergies. This is not as well

documented as the effect of probiotics on diarrhoea or a compromised immune system. Still, probiotics are a good option for parents of children with an increased risk of developing allergies or neurodermatitis. A large number of studies show they can offer significant protection. In some studies, the effect could not be proven, but that may be because each study used a different kind of bacterium. Personally, I think the 'better safe than sorry' approach is appropriate here. Probiotics certainly do no harm to children with a high risk of developing allergies. Some studies showed an improvement in the symptoms of those already suffering from allergies or neurodermatitis.

As well as the well-researched areas such as diarrhoea, gastrointestinal disease, and the immune system, there are other areas that are only now undergoing scientific scrutiny. Digestive complaints, traveller's diarrhoea, lactose intolerance, obesity, inflammatory joint disease, and diabetes are all promising areas of research.

If you ask your pharmacist to recommend a probiotic product to help with one of these problems (for example, constipation or flatulence), she will not be able to give you one that has been scientifically proven to work. The pharmaceutical industry and academic research are equally behind in this area. What remains for you is to try out different products for yourself, until you hit upon a bacterium that seems to help. The packaging should always include the name of the bacteria the product contains, so try it for about four weeks; if you see no improvement, give a different bacterium a go. Some gastroenterologists will advise you about the kind of bacteria that are more likely to be the ones you are looking for.

The rules are the same for all probiotics: you should try them for about four weeks, and make sure they are still within the best-before date (otherwise sufficient bacteria may not have survived to have any effect on the huge ecosystem of the gut). Before buying probiotic products, you should find out whether they are intended by the manufacturer for the complaint you are hoping to treat. Different bacteria have different genes — some are better at advising the immune system, others are more belligerent about ridding the gut of diarrhoea-causing bugs, and so on.

The best-researched probiotics to date are lactic acid bacteria (Lactobacilli and *Bifidobacteria*) and *Saccharomyces boulardii*, which is a yeast. It has not received the attention it deserves. It is not a bacterium, so it is not one of my favourites, either. But it does have one huge advantage: as a yeast, it has absolutely nothing to fear from antibiotics.

So, while we are massacring our entire bacteria population by taking antibiotics, *Saccharomyces* can move in and set up house without a worry in the world. It can then protect the gut from harmful opportunists. It also has the ability to bind toxins. However, it does cause more side effects than bacterial probiotics — some people have an intolerance to yeasts, which can cause them to break out in a rash, for example.

The fact that almost all the probiotics we know — give or take a yeast or two — are lactic acid bacteria shows how little we have yet discovered in this area. Lactobacilli are normally less-common residents in the guts of adults, and *Bifidobacteria* are unlikely to be the only health-promoting microbes present in the large intestine. At the time of writing, there is only one other probiotic bacteria species that is as well researched as the two mentioned above: *E. coli Nissle 1917*.

This strain of *E. coli* was first isolated from the faeces of a soldier returning from the Balkan War. All the soldier's comrades had suffered severe diarrhoea in the Balkans, but he had not. Since then, many studies have been carried out to show that this bacterium can help with diarrhoea, gastrointestinal disease, and a weakened immune system. Although the soldier died many years ago, scientists continue to breed his talented *E. coli* in medical laboratories and package it up for sale in pharmacies so it can work its wonders in other people's guts.

There is one limitation on the efficacy of all current probiotics we take: they are isolated species of bacteria bred in the lab. As soon as we stop taking them, they mostly disappear from our gut. Every gut is different, and contains regular teams that help each other or wage war on each other. When somebody new turns up, they are at the back of the queue when it comes to allocating places. So, currently, probiotics work like hair conditioner for the gut. When you stop taking them, the regular flora folk have to continue their work. To achieve longer-lasting results, researchers are now looking at the possibilities of a mixed-team strategy: taking several bacteria together, so that they can help each other gain a foothold in unknown territory. They clear away each other's waste, and produce food for their colleagues.

Some products you can buy at the pharmacy, drugstore, or supermarket already use this strategy, with a mix of our trusty old lactic acid friends. And it seems they really do work better as a team. The idea that we might be able to encourage these bacteria to settle permanently in our gut is a nice one, but has not yet been seen to work well ... to put it mildly.

But if you persist doggedly with the teamwork strategy, it can have impressive results — in treating *Clostridium difficile* infections, for example. *Clostridium difficile* is a bacterium that can survive treatment with antibiotics and then colonise the entire area left free by the bacteria killed by treatment. Infected people often have bloody, slimy diarrhoea for many years, which does not respond to further treatment with antibiotics or probiotics. This can put great strain on the body, but also the mind.

In difficult situations like this, doctors really have to use all their creativity. A few brave medics have now begun transplanting experienced teams of bacteria, including all possible real gut bacteria, from the guts of healthy donors to those with *Clostridium difficile* infections. Luckily, such a transplant is relatively easy to carry out (it has been used for decades by veterinary practitioners to cure many diseases). All you need is some healthy faeces, complete with bacteria, and that's it. The treatment is known as faecal bacteriotherapy, or, more directly, a stool transplant. The faeces used in medical stool transplants are not pure, but they are purified. Then it does not really matter whether it enters through the front or back door, so to speak.

Almost all studies show a success rate of around 90 per cent in treating previously incurable diarrhoea caused by *Clostridium difficile*. Few medical drugs have such a high success rate. Despite these positive results, the treatment can currently only be used on patients with truly hopeless cases of incurable diarrhoea. The danger is that other diseases or potentially harmful bacteria may be transplanted along with the donor stool. Some companies are already working to develop an artificial stool that they can guarantee is free of any harmful elements. If they succeed, the therapy is likely to become much more widespread.

Probiotics' greatest potential probably lies in transplanting beneficial bacteria to the gut, where they would settle permanently and grow. Such transplants have already helped patients with severe diabetes. Scientists are currently investigating whether they can also be used to prevent patients from ever developing full-blown type 1 diabetes.

The connection between a stool and diabetes may not be immediately obvious to everyone. But, in fact, it is not as absurd as it seems. It is not just defensive bacteria that are transplanted with the stool, but an entire microbial organ that plays an important part in regulating the body's metabolism and immune system. We are still completely ignorant of more than 60 per cent of these gut bacteria. Looking for species that may have probiotic effects is time consuming and difficult, just as the search for medicinal plants was in the past. Only this time, the search is going on inside us. Every day, and every meal we eat, influences the great microbial organ inside us — for better or for worse.

Prebiotics

That is the central idea surrounding the use of prebiotics — supporting our good bacteria by eating certain foods. Prebiotics are much more suitable for daily use than probiotics. To gain the benefits they offer, just one condition must be met: good bacteria must already be present in the gut. These can then be encouraged by eating prebiotic food, which gives them more power over any bad bacteria that may also be present.

Since bacteria are so much smaller than we are, they view food from a very different perspective from our own. Every little grain becomes a major event, a comet of deliciousness. Food we cannot digest in the small intestine is called dietary fibre or roughage. But, despite the name, it is not rough on the bacteria of the large intestine. Quite the opposite, in fact: they love it! Not all kinds, but some, anyway. Some bacteria love undigested asparagus fibres; others prefer undigested meat fibres.

Some doctors who recommend their patients eat more dietary fibre do not even really know the reason. They are prescribing a hearty meal for your bacteria that will benefit you, too. Finally, your gut microbes get enough to eat, so they can produce vitamins and healthy fatty acids, or put the immune system through a good training session. However, there are always some pathogens among the bacteria in our gut. They can use certain foodstuffs to produce substances such as indole, phenols, or ammonia, for example. Those are the substances you find in the chemistry cupboard with a warning symbol on the bottle.

This is exactly how prebiotics can help: they are roughage that can only be eaten by nice bacteria. If such a food were available in the human world, the works canteen would be quite an eye-opener!

Household sugar is not a prebiotic, for example, because it is also a favourite of tooth-decay bacteria. Bad bacteria cannot process prebiotics at all, or hardly, and so they cannot use prebiotics to produce their evil chemicals. At the same time, good bacteria fed with prebiotics grow constantly in power, and can gain the upper hand in the gut.

And it is not even very hard to do yourself and your best microbes a favour with prebiotics. Most people already have a favourite prebiotic dish that they would not mind eating more often. My Granny always has some potato salad in her fridge, my Dad's speciality is a great endive salad with mandarin segments (here's a tip: rinse endive leaves briefly under warm water — this leaves them crispy, but removes some of the bitterness), and my sister can't resist asparagus or black salsify in a creamy sauce.

Those are just a few dishes that *Bifidobacteria* and Lactobacilli love to eat. We now know that they prefer liliaceous vegetables — those from the lily family (*Liliaceae*) — which include leeks and asparagus, and onions and garlic. They also like *Compositae*-plants, which are those from the sunflower family, including endives, salsify, artichokes, and Jerusalem artichokes. Resistant starches are also on their favourites list.

Resistant starches form, for example, when potatoes or rice are boiled and then left to cool. This allows the starch to crystalise, making it more resistant to digestion. So more of your potato salad or cold sushi rice reaches your microbes untouched. If you don't already have a favourite prebiotic dish, give them a try. Eating these dishes regularly has an interesting side effect: it causes regular cravings for just such foods.

People who eat mainly low-fibre foods such pasta, white bread, or pizza should not suddenly switch to eating large portions of

---

Fig.: *Artichoke, asparagus, endive, green bananas, Jerusalem artichoke, garlic, onion, parsnip, black salsify, wheat (wholegrain), rye, oats, leek*

high-fibre foods. That will only overwhelm their underfed bacterial community. The sudden change will freak them out, and they will metabolise everything they can in a fit of euphoria. The result is a never-ending trumpet concerto from down below. So, the best strategy is to gradually increase the amount of dietary fibre, and not to feed your bacteria with massive, unmanageable amounts. After all, the food we eat is still first and foremost for us, and only secondarily for the inhabitants of our large intestine.

Overproduction of gas is not a pleasant thing: it bloats the gut, making us feel uncomfortable. But passing a bit of wind is not only necessary; it is healthy, too. We are living creatures, with a miniature world living inside us, working away and producing many things. Just as we release exhaust fumes into the earth's atmosphere, so must our microbes, too. It may make a funny sound, it may smell a bit — but not necessarily. *Bifidobacteria* and Lactobacilli, for instance, do not produce any unpleasant odours. People who never need to break wind are starving their gut bacteria and are not good hosts for their microbe guests.

Pure prebiotics can be bought at the pharmacy or drugstore. This include the prebiotic *inulin*, which is extracted from the roots of the endive plant, and GOS (*galactooligosaccharides*), which are isolated from milk. These substances have been tested for their health-giving effects, and are pretty efficient at feeding only certain *Bifidobacteria* and Lactobacilli.

Prebiotics are nowhere near as well researched as probiotics, although there are already some sound results pertaining to how they work. Prebiotics promote good bacteria, resulting in a reduction in the amount of toxins produced in the gut. People with liver problems, especially, are unable to detoxify these substances well, and that can cause serious problems. Bacteria toxins have various effects on the body, including anything from fatigue and tremors to comas. When such patients are treated in hospital, they are often given highly concentrated prebiotics, which usually leads to a disappearance of their symptoms.

But bacterial toxins also influence Joe Bloggs with his fully functioning liver. They can develop, for example, if he eats too little dietary fibre and it is all used up at the beginning of his large intestine. The bacteria at the end of his gut will then pounce on any undigested proteins. Bacteria and meat can be a bad combination — we know it's never a good idea to eat rotten meat. Too many meat toxins can damage the large intestine and, in a worst-case scenario, can even cause cancer. The end of the gut is more prone to cancers on average than the rest of the organ. That's why researchers are so keen to test how well prebiotics can protect against cancer. Early results are promising.

Prebiotics like GOS are interesting, because they are also produced by our own bodies. Ninety per cent of the roughage in human breast milk is GOS; the remaining 10 per cent is made up of other indigestible fibre. In cow's milk, GOS accounts for only 10 per cent of the fibre content. So it appears there is something about GOS that is important for human babies. If bottle-fed babies are given formula milk that contains a little GOS powder, their gut bacteria look similar to those of breastfed babies. Some studies indicate that they are less prone to allergies and neurodermatitis in later life than other bottle-fed babies. GOS has been approved as an additive to baby-milk formula since 2005 — but it is not obligatory.

Since then, interest in GOS has increased, and another of its effects has now been demonstrated in the laboratory. GOS docks onto gut cells directly — preferably at locations otherwise favoured by pathogens. That means they act as microscopic shields. Bad bacteria cannot get a good hold, and are more likely to slip and fall away. These results have prompted new studies into GOS as a way of preventing traveller's diarrhoea.

More research has been carried out into *inulin* than GOS. It is sometimes used as a sugar or fat substitute in the food industry, because it is a little bit sweet and has a gel-like consistency. Most prebiotics are sugars that are linked into chains. When we speak of sugar, we often mean a particular molecule extracted from sugar beet — but there are more than a hundred different kinds of sugar. If, historically, we had developed a sugar industry based on endive sugar, our sweets would not cause tooth decay. 'Sweetness' is not in itself unhealthy; we simply eat only the most unhealthy kind of sweetness.

We are often sceptical about products that are labelled 'sugar-free' or 'low fat'. Sweeteners such as aspartame appear to be carcinogenic, while other substances used in typical 'diet' products are also used in factory farming to fatten pigs. So our scepticism is justified. But a product that contains *inulin* as a sugar or fat substitute may well be healthier than one with a full dose of animal fat or sugar. It is worth reading the label on diet products closely, because we can consume some of them with a clean conscience, and our gut microbes will thank us for it, too.

Inulin cannot bind with our cells as well as GOS. A very large and well-run study in the UK showed that it did not offer any protection against traveller's diarrhoea — although the test subjects reported a general improvement in wellbeing after taking *inulin*. This pleasant effect was not reported by members of the control group, who were given a placebo. *Inulin* can be produced with different chain lengths, which is great for attaining a beneficial distribution of good bacteria. Short *inulin* chains are eaten by the bacteria at the start of the large intestine; longer chains are consumed closer to the end.

This so-called ITF_{MIX} containing chains of differing lengths produces good results, where more surface area is beneficial — in the absorption of calcium, for example, which relies on bacteria to help it pass through the gut wall. In one experiment, ITF_{MIX} was seen to improve calcium absorption in young girls by up to 20 per cent. That is good for the bones, and can help protect against osteoporosis in old age.

Calcium is such a good example because it shows the limits of what can be achieved with prebiotics. First, you have to ingest enough calcium for an effect to kick in; and, second, prebiotics can do nothing if the problem lies with other organs. When they go through menopause, many women's bones get weaker. This is due to a mid-life crisis in the ovaries. They have to say goodbye to their life of producing hormones and learn how to enjoy their retirement. But the bones miss those hormones. No prebiotic in the world can help with this kind of osteoporosis.

Nevertheless, we should not underestimate the power of prebiotics. Almost nothing influences our gut bacteria as much as the food we eat. Prebiotics are the most powerful tool at our disposal if we want to support our good bacteria — that is, those that are already there and are there to stay. Prebiotic creatures of habit, like my potato-salad-addicted Granny, are doing the best part of their microbial organ a favour, without even knowing it. Incidentally, her second-favourite vegetable is leeks. Whenever everyone else was ill, she would be there with a big smile and some soup, merrily playing us a few songs on the piano. I don't know if her defences had anything to do her microbes, but it's not an illogical conclusion.

So, we should remember: good bacteria do us good. We should feed them well so they can populate as much of our large intestine as possible. Pasta and bread made of white flour on factory production lines are not enough. We need to include real roughage, made of real dietary fibre in vegetables and fruit. They can also satisfy a sweet tooth, and taste delicious. It can be fresh asparagus, sushi rice, or pure, isolated prebiotics from the pharmacy. Our bacteria will like it, and they will thank us with their good services.

Seen under the microscope, bacteria look like nothing but little bright spots against a dark background. But, taken together, their sum is much greater than their parts. Each one of us hosts an entire population. Most sit in our mucous membrane, diligently training our immune system, soothing our villi, eating what we don't need, and producing vitamins for us. Others keep close to the cells of the gut, needling them or producing toxins. If the good and the bad are in equilibrium, the bad ones can make us stronger, and the good ones can take care of us and keep us healthy.

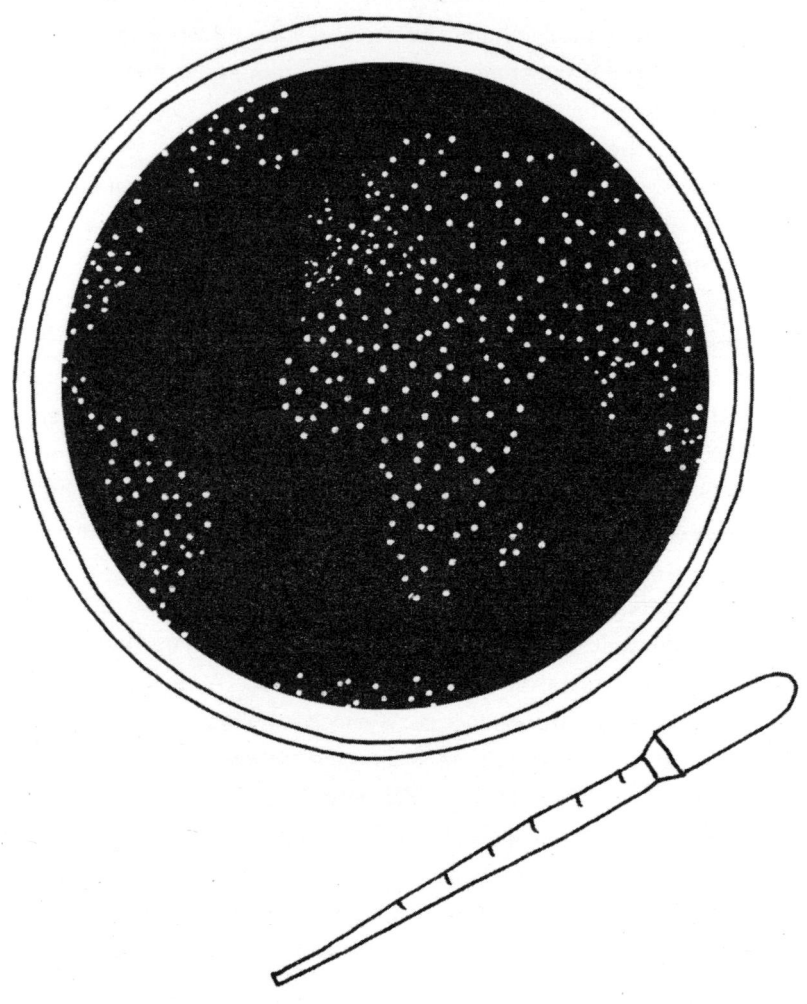

The Gut-Brain Axis

When I first sat down in 2013 to write about the gut-brain axis, I spent an entire month staring at a blank screen. At that time, the gut-brain axis was still an extremely new field of research — nearly all the studies carried out till then had been on animals, and so the area was more one of speculation and hypotheses than hard facts. I was determined to include information about the experiments and theories that existed; but, at the same time, I was anxious to avoid raising false hopes or spreading half-truths by jumping the academic gun, so to speak. As I sat sniffling at my sister's kitchen table one grey Thursday, moaning that I wouldn't be able to come up with a text that was both accurate and clear enough, she eventually told me, almost as a command, 'Get your head down and just write up what you understand of it, and if there's more or better information in a few years' time, I'm sure you can add it then.'

And that's exactly what I've done.

New discoveries

Doing academic research is like walking through unfamiliar territory in the fog. There are few people who would wish to do that every day, happily snapping photos of the new bushes or house walls that emerge from the mist. Sometimes, you can follow a woollen thread for ages, only to find that it is your own jumper that is unravelling. And then coming home and telling your loved ones what discoveries you've made is only...so-so cool.

A couple of years ago, we knew that some depressed mice could be cheered up by certain bacteria, and we knew of rats whose character changed completely when other rats' gut bacteria were transplanted into their bodies, and thus the term 'psychobiotics' was coined. It describes microbes that have psychological effects — and which may even be useful in treating conditions like depression. But those animal studies still left us wondering whether such psychobiotics could actually affect humans' mental health.

There are now some twenty reliable studies of this kind involving human subjects. Three of the bacteria cocktails tested appeared to have no such effect, but all the rest (and this is the more exciting piece of news) do influence the human psyche. Overall, a realistic view of the picture so far is that bacteria don't cause sudden mood swings; rather, their effect on our emotional state is slow-working, often not appearing until three to four weeks after the microbes are administered, and their effect may be fairly limited. Medical thought has also undergone a development when it comes to stress, and the gut has been recognised as playing an important role.

What exactly is going on here, and how strong the effects

might be, are still under investigation, with nearly every research team concentrating on different kinds of bacteria. metabolism.

Mood

For researchers conducting experiments into our mood, a stimulating initial question is: What emotions constitute our moods? What are they made up of, so to speak? Most researchers try to answer this with the aid of questionnaires. The many different kinds of questions are categorised according to ranges of feelings: from depressive to cheerful, from fearful to self-assured, from quick-tempered to sweet-tempered, concerned (for example about personal physical wellbeing), benevolent, and so on.

And that is how a group of English researchers began their first, tentative trials. *Lactobacillus casei* Shirota (a bacterium you may know from the yoghurt drinks available at most supermarkets) improved the disposition of the most ill-tempered third of their test subjects after they had taken it for only three weeks, from 'depressive' towards 'cheerful'. The subjects with a better general mood saw no further improvement in their spirits, and other feelings, such as anger or anxiety, were not affected to any real degree.

That was not the case for a study carried out in France. Scientists wanted to investigate a combination of two types of bacteria (*Bifidobacterium longum* and *Lactobacillus helveticus*, which are often found in their abbreviated forms, *B. longum* and *L. helveticus*, in the 'list of ingredients'). After four weeks, the test subjects experienced a positive improvement not only in their depressive tendencies, but also in parameters such as anger; or, for example, they showed a propensity to perceive their own physical aches and pains as less severe.

A Dutch team took a more specific approach to the category of 'mood' itself. They were particularly interested in a certain kind of mood: those minor, recurrent, routine lows familiar to everybody, even those of a perfectly healthy mental constitution. Nothing terrible has happened, and we really can't even pinpoint what's causing our bad mood — we just feel less perky than usual. Such moments of gloom rarely warrant a mention in our day-to-day conversations or in the media — but they are currently a hot topic among research psychologists. What the researchers are interested in is not the blues themselves, but how we react to them.

This is among the most accurate indicators of whether a healthy person might develop depression or not. Above all, studies have repeatedly shown that the least favourable reaction is fretful brooding — about who might be to blame for our woes, for example.

In this study, the test subjects were asked to spend several minutes imagining themselves in the following situation: It's definitely not your day today, you don't feel in top form, but nothing serious has happened to put you in this mood. The subjects then evaluated their own reaction to this mood, according to statements such as:

'When I feel that way, I have much less patience and I lose my temper more quickly.'

'When I feel that way, I brood a lot over everything that's wrong with my life at the moment, or about what my life might have been like if it had taken a different turn.'

Or also:

'When I'm down, I feel that everything is hopeless.'

In the fine old tradition of test questionnaires, the subjects were given the options

'Nope, never really / rarely / sometimes / often / oh yes — absolutely!' (0 to 4 points).

Before they began taking the bacteria, the test subjects' average scores were about 43 out of a total of 136 points. Thus, they were within the healthy average range and did not display unusually strong brooding, angry, or despairing reactions. They then took the bacteria mixture every day for four weeks. Mouth open, powder in, swallow. The group that were given a placebo (without their knowledge) barely changed their responses at all. But those who took the real bacteria improved their scores — in particular, in two areas, anger and brooding — by no less than about 10 per cent (which means roughly half the questions in those areas were answered one degree more positively).

This is not the same effect as would have been achieved if they had been given cocaine — or a powerful tranquiliser — but it was also not the same effect as that of the placebo. Results like these lead us to the question of how great the gut's influence on our mood is. They also beg the question of what aspects of our mood it can affect.

Stress

While mood can be seen as originating in various parts of our nervous system, stress is better described as the state the nervous system is in. A stressed nervous system is like a taut bowstring — in a constant state of alertness, and sensitive to any external stimulus. Such a state is excellent for tackling an obstacle course or reacting to a dangerous situation while driving. However, as a permanent lifestyle, such a state is pretty costly... comparable to taking a monster truck to drive to the supermarket around the corner. The gut lends the brain a large amount of energy in order to deal with the stress (see p. 124). Could our tummies also help relieve feelings of stress? As a way of helping themselves, so to speak?

In this field, the wait for newer results has been worth it, as the research has changed tack. The initial conclusion following the first experiments with human subjects was that a stressful daily routine or a scary exam were always equally stressful or frightening for subjects, no matter which lovely bacteria they fed to their guts. However, it was concluded, microbes could help reduce the physical effects of stress, such as stress hormone levels, nervous tummy aches, nausea, diarrhoea, and susceptibility to colds.

However, a closer examination of those studies reveals an interesting indication. One species of bacteria also altered people's subjective level of stress — although this appeared to be the case only in one sub-group of subjects: the sleep-deprived ones. Those who slept less in the run-up to an exam were always more stressed. This effect was less pronounced among subjects who swallowed a daily dose of *Bifidobacterium bifidum*. They were still stressed — but a little less so than their peers who slept equally badly but did not get the bacterium. Two other species of bacteria (*Lactobacillus helveticus* and *Bifidobacterium infantis*) were also tested in these trials, but were not found to have the same effect.*

This result encouraged scientists to investigate this microbial world in more depth — after all, a bacterium had now been found that managed something none other had been shown to do: reduce feelings of stress. It was not long before new results emerged from Ireland. Some researchers — those who had also carried out the experiment with the swimming mice (see p. 118) — now ventured into the arena of human trials. One particular Bifidobacterium (*Bifidobacterium longum 1714*) had shown itself to be highly effective in their experiments with mice. It reduced stress parameters and improved the lab animals' memory. So the Irish scientists tested it on a small group of human animals.

The participants in this study were asked to fill out an online questionnaire about their feelings of stress every day. They were

---

* In this study, 581 subjects were divided into four groups a couple of weeks before it began. One group received capsules containing a powder with no effect, the other three groups were each given a different type of bacteria (*Lactobacillus helveticus*, *Bifidobacterium infantis* or *Bifidobacterium bifidum*).

also called into the lab three times within a period of eight weeks, in order to:

1. put on a funny helmet;
2. plunge one of their hands into ice-cold water;;
3. solve mental puzzles on a tablet PC.

The function of the funny helmet was to measure the activity levels in various parts of the subjects' brains. If you were to make your beloved wear this helmet while you were telling him or her about your boring day at the office, you would be able to watch as the level of activity in the listening area of their brain slowly declined and the day-dreaming regions revved up for a party.

Plunging subjects' hands into unbearably cold water is a tried-and-tested method of assessing their stress levels. A measurement is taken of the amount of time subjects are able to keep their hands in the icy water, while swabbing their mouths regularly with cotton buds. The cotton buds absorb saliva, which can be analysed to measure the amount of stress hormones it contains. No matter how often the test is repeated, the subjects' reactions are always the same. A nervous system that signals stress due to cold is not a creature of habit — it does not get less sensitive over time when exposed to low temperatures repeatedly or for a long period of time. If it did, we would perceive the weather as getting increasingly warmer as winter progressed.

As soon as the test subjects had had enough and rescued their hands from their freezing fate, they were asked a number of questions. The researchers wanted to know how anxious they felt immediately after experiencing a rush of excited hormones.

Almost every parameter was different in some way after a four-week course of Bifidobacteria. In the online questionnaires, the test subjects reported around 15 per cent lower levels of day-to-day stress, as compared with the subjects who received a placebo. The cold-water hand-immersion test still triggered the same reaction (as expected — after all, the water was still freezing cold), but with an overall stress hormone level that was lower than before. Furthermore, those hormones also did not lead to increased anxiety.

The electrode helmet and tablet-based teasers were also not simply for the researchers' amusement. The 'bacteria group' made around two to five fewer mistakes in the memory tasks, compared to the placebo group (who made one to three fewer mistakes) — already a successful result. This effect was also made visible by the helmet. One area of the brain that we use for learning, and which becomes weaker in patients with Alzheimer's, now showed increased activity. The placebo had no such effect — but with the bacteria, it was clear.

Our gut can send impulses to the brain — for example, via nerve fibres (see p. 118) The Irish research team came up with another possible explanation — bacteria could have improved memory function by reducing the subjects' stress hormone levels. That might work like this: the brain structure which stores our memories and links them to each other (the hippocampus) is very densely populated with sensors that detect stress hormones. If the hippocampus registers large amounts of such hormones, the brain cuts back the level of activity there. After all, if you are running away from a wild animal, you don't need to waste energy remembering which plants you pass by. During stressful times in our lives we develop a kind of tunnel vision — to enable us to direct our attention to the problem at hand.

This observation is interesting — not only for fans of *Bifidobacterium longum 1714*, but also for patients with gut conditions who find it difficult to concentrate as well as usual during a flare-up, and for school students, for example, who have trouble cooperating with their brains during a tough test. It may well be the case that it is not always so important whether the state of stress is reported by the gut or the brain. Both organs are able to use the nervous system and messenger substances in the blood to stimulate the adrenal gland (the organ that eventually produces stress hormones). And this is precisely where we may return to the issue of mood.

Let's go back to those little, day-to-day downers studied by the group of Dutch scientists. Remember — no matter how nice our lives may be, we all have them from time to time. Those who tend to react to such a mood by brooding over their problems will find that our modern world offers a rich palette of problems to ponder. Even more importantly, these are often problems we can do nothing about. Some politician somewhere in the world says something stupid, but in the past hardly anyone would have got wind of it. Elsewhere, a plane comes down, killing an entire football team you would never have otherwise heard about. Someone or other puts a picture of her perfect life online, and we end up comparing ourselves unfavourably to something we would never have seen in the past.

If we engage in it enough, brooding over problems that cannot be changed can result in feelings of stress. The resultant stress hormones serve to increase that tunnel vision. And it becomes increasingly difficult for us to see anything beyond our own problems. And this, in turn, further increases our level of stress. *Voilà*: we have a vicious circle. A physiological system that was meant to help us in times of stress is thus co-opted, and we increasingly slide into a behaviour pattern of 'stressed griping', rather than directly observing the world around us, asking inquisitive questions or taking care of our wellbeing.

Depression

The current state of research indicates that our gut has about a 10-to-15 per cent influence on feelings of melancholy, anger, or stress. It tells the brain what is happening inside us, and that information may be worrying or reassuring. So it could be partly responsible when we slide into a certain mood. However, this says nothing about the extent of the gut's involvement in the process of emerging from a period of fully fledged depression.

Preliminary experiments aimed at answering this question give cause for optimism. A group of Irish researchers, for example, harvested gut bacteria from people with depression, and implanted them in rats. This is not the kind of microbial exchange that happens when you shake hands with someone: the scientists first removed all other microbes from the rats' intestinal system and then administered highly concentrated doses of the harvested bacteria. The rats developed depressive behaviours that they had not displayed previously.

When it comes to purely human experiments, research is still at a very early stage. A common tool in such research is the 'Becks Depression Inventory' psychometric test. This test helps scientists determine the severity of a patient's depression (and, for example, whether they are dealing with a case of clinical depression or a temporary bout of the blues). Perhaps surprisingly, the list of 21 questions does not only ask respondents whether they feel sad or dissatisfied; it also includes questions about problems with sleeping, making decisions, increasing health worries, or a noticeable lack of interest in sex (compared to the past). Basically, these are indirect questions about the various hormone systems in our body.

So far, there are only two studies that have investigated, under controlled conditions, the effect of probiotic bacteria on depression. The first study, in 2015, found that a combination of two kinds of bacteria (*Lactobacillus acidophilus* and *Bifidobacterium bifidum*) and medication improved patients' condition, but the effect was slight, once all possible interference factors were removed from the calculations. And a study in 2017 found that two types of bacteria (*Lactobacillus helveticus* and *Bifidobacterium longum*) had no influence on depression. However, the researchers did find an indication that a patient's vitamin D level may influence this effect. In test subjects with sufficiently high levels of vitamin D in their blood, the microbes appeared to improve their mood; however, the overall number of subjects was too small to allow a scientific conclusion from this study.

These are the first two steps on a new journey in research. If we continue down this path, we will eventually see which way it is leading us. Could we perhaps use the gut to prevent depression before it occurs? Are bacteria better suited as a complementary treatment, to be used alongside medication, therapy, and lifestyle changes? Or should treatments for depression perhaps target multiple possible causes at once — that is, the gut (via bacteria and diet), the brain (with drugs and therapy), and other possible causes (such as vitamin levels, physical activity, or working conditions)? It may also be the case that there are many different kinds of depression — some for which the influence of the gut is greater and therefore more important in their treatment, and others for which this is not the case.

The goal of this journey should not be to discover some kind of 'super-bug', which we would all take a daily dose of. The aim is also not to produce a population of permanently happy people. Rather, the aim should be to reach a better understanding of our bodies and how we live in them. This includes being aware of what is going on inside our bodies, rather than looking only for external causes of stress or mood swings. If we were to discover particularly effective microbes to support us along the way, that would be an excellent thing. But while scientists are still searching through the fog, we should learn to appreciate what we already have: good microbes inside us and around us, and ancient knowledge, to which we should pay much more heed.

Clever cravings for fermented foods

Sometimes it is not necessary to understand our desires and cravings. If a person occasionally enjoys rolling around on their living-room floor for no apparent reason, let them. However, it can sometimes be both useful and interesting to find out why we develop the desire to eat and drink what we do.

Take a glass of water, and add an equal amount of sugar to that contained in the same volume of Coke. Few people would want to drink the resulting mixture. And if we did, we would probably feel disgust at the idea of downing a second glass. This is because our bodies are cleverer than we think.

If we now do something that several million years of evolution has not been able to prepare us for, the experiment ends with a very different result. If we add a little citric acid to the sugar water (represented by carbonic and phosphoric acid in Coke) — hey presto! We have a delicious drink. We down the glass in one gulp, and our brain claps its hands in delight: hooray!

Our bodies are familiar with acid, from fruit and good bacteria (for example, the lactic acid bacteria in yoghurt). When the acid is not too strong and comes in combination with other nutrients, it gains the trust of our taste buds. Thus we like to include an acid component when we cook a pleasant-tasting meal — tomatoes in the sauce, a squeeze of lemon over the fish, a glug of wine over the frying onions. This well-placed pleasure is fascinating for anyone with an interest in microbes. A pertinent question might be: when we fancy a sour ingredient in our meal, are we really craving good bacteria?

Down the millennia of human history, anyone with a yen for

a taste of 'sour' would satisfy their hunger with good microbes. Our forebears fermented cabbage to create sauerkraut, and drank wine rather than water (which was usually too contaminated to drink untreated in the Middle Ages). They still baked their bread using real sourdough, and made their own sour milk and yoghurt products. They did not have access to citrus fruits or acidified fizzy drinks. This observation gives cause for a little experiment you can carry out yourself.

Fermenting vegetables with bacteria at home — a.k.a making sauerkraut

Fermenting means getting bacteria to pre-digest your food. Bad bacteria and moulds do not ferment your food nicely, but spoil it and render it inedible. Good bacteria, however, process our food to make it easier for us to digest. They are better than our digestive enzymes at splitting open cabbage cells (or other plant cells). In this way, they make the work of the gut much easier, and even produce additional vitamins in the process. They also produce acids that kill off any dangerous bacteria, thus preserving the food for longer. Since good bacteria are everywhere, just hanging around in the environment, it makes sense to provide them with a useful job to do and something to eat. This helps them multiply and gain more power.

1. Cabbage is the classic, but almost any vegetable that can be eaten raw can also be fermented. Carrots and gherkins, for example (gherkins are the pinnacle of pickling, however, since precise procedures must be followed if they are to remain crisp to the bite). Some good bacteria will already be on the cabbage leaves or the skin of the carrots, so there is no need to add any particular bacteria to them. This is why it can be beneficial to buy vegetables that have not been treated with pesticides, etc.

2. Depending on how long you want the fermentation process to take, slice your vegetables thinly or grate them (= one week's fermentation time) or leave them whole (= four to six weeks' fermentation time). You need to make sure you avoid contamination as you work. You don't want just any old kitchen bacteria joining your product in the jar.
3. Add 10 to 15 grams of salt for every kilogramme of vegetables. This slows the growth of bacteria in general, preventing bad germs from commandeering the process before the good ones can do their work. It is important to add the correct amount of salt: too much will prevent the fermentation process altogether; too little can result in the food going off and tasting bad. Sea salt is a good choice, but do not use iodised salt, as the iodine inhibits the bacteria's growth too much.
4. Knead the mixture with appropriate fervour. This is to mix the salt well into the concoction, and helps to partially break down particularly tough cell walls. The salt helps extract the water from the cabbage cells, which can be used later as the pickling liquid.
5. Press the cabbage firmly into a jar with an air-tight seal. It is important to make sure that all the vegetable matter is submerged beneath the liquid so that it is not exposed to oxygen, which disturbs the fermenting bacteria. Any parts that stick out above the liquid will not be protected by the acid, and could go mouldy. If your cabbage or carrots have not produced enough liquid of their own to cover them completely, you can add more salted water (add a generous teaspoon of salt for every 250 ml of water). If there is still some vegetable matter sticking out above the surface of the liquid, you can add a weight to press it down. (Special 'sauerkraut weights' are available, in Germany, at least! — but a rock or stone of the right size, which has been cleaned by boiling it in water, will do the job just as well). Those who prefer a particular flavour to

their sauerkraut can now add almost anything they want to the jar, such as caraway seeds, beetroot, or, particularly good with carrots, a little ginger.

You may sometimes see little bubbles of gas rising through the liquid as the fermentation process progresses. This is why people in ancient cultures, who had no knowledge of the microbial world, would sometimes dance around the fermentation vats, as they thought this would encourage the vegetables to bubble. Others, by contrast, would afford the barrels calm so as not to disturb the gods in their work. Once fermentation is complete and everything is ready, the result should taste sour, and should not be too slimy in consistency or taste too strongly of alcohol. From this stage, it can be kept in the fridge.

Incidentally, it is at this stage that the sauerkraut you buy from the supermarket is boiled to pasteurise it. This not only kills off the bacteria, but it also destroys some of the vitamin C they have produced. Manufacturers therefore often add vitamin C powder after pasteurisation. The acid it relies on means that fermentation is the safest way to preserve food — canned or bottled foods have been known to cause illness due to temperature-resistant bacteria, but no case of illness caused by fermented foods has ever been reported.

You can now add a spoon or two of your home-made sauerkraut or sour carrots to just about any meal: in salad instead of vinegar, on a burger instead of gherkin pickled in vinegar, in soups and stews (add just before serving), with vegetable or rice dishes, or, for those feeling particularly adventurous, with some honey in your breakfast porridge. Then just wait and see if you start craving this special 'sourness' more and more. When your appetite decides it likes something healthy after trying it once or twice — you might as well go with it.

Acknowledgements

This book would not exist if it weren't for my sister Jill. Without your free, rational, and inquisitive mind, I would have been stuck many times in a world where obedience and conformity are easier than courage and the will to make necessary mistakes. Although you lead a busy life yourself, you were always there, ready to go through my texts with me and inspire me with new ideas. You taught me how to work creatively. Whenever I feel bad, I remember we are made of the same stuff, and each of us uses her talents differently.

I would like to thank Ambrosius, who shielded me from too much work with a protective arm. I would also like to thank my family and my godfather, for surrounding me like a forest surrounds a tree, keeping me rooted even when strong winds are blowing. I also thank Ji-Won, for keeping me nourished so often during the writing of this book — with food and her wonderful nature. My thanks go to Anne-Claire and Anne for their help with even the trickiest of questions.

I thank Michaela and Bettina, without whose sharp minds the writing of this book would never have come about. If it weren't for my medical studies I would never have had the knowledge necessary to write this book, so I thank all my good teachers and professors, as well as the German state, which pays for my university studies. To everyone who has contributed their hard work to the realisation of this book — from press officers, publishers, manufacturers, printers, marketers, proofreaders, booksellers, and postal workers, to those of you reading this now: Thank you!

Main References

This list mainly contains references to sources dealing with content not normally covered in the standard textbooks.

Part One

Bandani, A.R.: 'Effect of Plant a-Amylase Inhibitors on Sunn Pest, Eurygaster Integriceps Puton (Hemiptera: Scutelleridae), Alpha-Amylase Activity'. In: *Commun Agric Appl Biol Sci.* 2005; 70 (4): pp. 869–73.

Baugh, R.F. et al.: 'Clinical Practice Guideline: Tonsillectomy in Children'. In: *Otolaryngol Head Neck Surg.* 2011 January; 144 (Suppl. 1): pp. 1–30.

Bengmark, S.: 'Integrative Medicine and Human Health — The Role of Pre- Pro- and Synbiotics'. In: *Clin Transl Med.* 2012 May 28; 1 (1): p. 6.

Bernardo, D. et al.: 'Is Gliadin Really Safe for Non-Coeliac Individuals? Production of Interleukin 15 in Biopsy Culture from Non-Coeliac Individuals Challenged with Gliadin Peptides'. In: *Gut.* 2007 June; 56 (6): p. 889 f.

Bodinier, M. et al.: 'Intestinal Translocation Capabilities of Wheat Allergens Using the Caco-2 Cell Line'. In: *J Agric Food Chem.* 2007 May 30; 55 (11): pp. 4576–83.

Bollinger, R. et al.: 'Biofilms in the Large Bowel Suggest an Apparent Function of the Human Vermiform Appendix'. In: *J Theor Biol.* 2007 December 21; 249 (4): pp. 826–31.

Catassi, C. et al.: 'Non-Celiac Gluten Sensitivity: The New Frontier of Gluten Related Disorders'. In: *Nutrients*. 2013 September 26; 5 (10): pp. 3839–53.

Kim, B.H.; Gadd, G.M.: *Bacterial Physiology and Metabolism*. Cambridge: Cambridge University Press, 2008.

Klausner, A.G. et al.: 'Behavioral Modification of Colonic Function. Can Constipation Be Learned?'. In: *Dig Dis Sci*. 1990 October; 35 (10): pp. 1271–75.

Lammers, K.L. et al.: 'Gliadin Induces an Increase in Intestinal Permeability and Zonulin Release by Binding to the Chemokine Receptor CXCR3'. In: *Gastroenterology*. 2008 July; 135 (1): pp. 194–204.

Ledochowski, M. et al.: 'Fructose- and Sorbitol-Reduced Diet Improves Mood and Gastrointestinal Disturbances in Fructose Malabsorbers'. In: *Scand J Gastroenterol*. 2000 October; 35 (10): pp. 1048–52.

Lewis, S.J.; Heaton, K.W.: 'Stool Form Scale as a Useful Guide to Intestinal Transit Time'. In: *Scand J Gastroenterol*. 1997 September; 32 (9): pp. 920–24.

Martín-Peláez, S. et al.: 'Health Effects of Olive Oil Polyphenols: Recent Advances and Possibilities for the Use of Health Claims'. In: *Mol. Nutr. Food Res*. 2013; 57 (5): pp. 760–71.

Paul, S.: *Paläopower — Das Wissen der Evolution nutzen für Ernährung, Gesundheit und Genuss*. Munich: C.H. Beck-Verlag, 2013 (2nd Edition). (In German)

Sikirov, D.: 'Etiology and Pathogenesis of Diverticulosis Coli: A New Approach'. In: *Med Hypotheses*. 1988 May; 26 (1): pp. 17–20.

Sikirov, D.: 'Comparison of Straining During Defecation in Three Positions: Results and Implications for Human Health'. In: *Dig Dis Sci*. 2003 July; 48 (7): pp. 1201–05.

Thorleifsdottir, R.H. et al.: 'Improvement of Psoriasis after Tonsillectomy Is Associated with a Decrease in the Frequency of Circulating T Cells That Recognize Streptococcal Determinants and Homologous Skin Determinants'. In:

J Immunol. 2012; 188 (10): pp. 5160–65.

Varea, V. et al.: 'Malabsorption of Carbohydrates and Depression in Children and Adolescents'. In: *J Pediatr Gastroenterol Nutr.* 2005 May; 40 (5): pp. 561–65.

Wisner, A. et al.: 'Human Opiorphin, a Natural Antinociceptive Modulator of Opioid-Dependent Pathways'. In: *Proc Natl Acad Sci USA.* 2006 November 21; 103 (47): pp. 17979–84.

Part Two

Aguilera, M. et al.: 'Stress and Antibiotics Alter Luminal and Wall-adhered Microbiota and Enhance the Local Expression of Visceral Sensory-Related Systems in Mice'. In: *Neurogastroenterol Motil.* 2013 August; 25 (8): pp. e515–e529.

Bercik, P. et al.: 'The Intestinal Microbiota Affect Central Levels of Brain-Derived Neurotropic Factor and Behavior in Mice'. In: *Gastroenterology.* 2011 August; 141 (2): pp. 599–609.

Bravo, J.A. et al.: 'Ingestion of *Lactobacillus* Strain Regulates Emotional Behavior and Central GABA Receptor Expression in a Mouse via the Vagus Nerve'. In: *Proc Natl Acad Sci USA.* 2011 September 20; 108 (38) pp. 16050–55.

Bubenzer, R.H.; Kaden, M.: at www.sodbrennen-welt.de (Retrieved October 2013)

Castrén, E.: 'Neuronal Network Plasticity and Recovery from Depression'. In: *JAMA Psychiatry.* 2013: 70 (9): pp. 983–89.

Craig, A.D.: 'How Do You Feel — Now? The Anterior Insula and Human Awareness'. In: *Nat Rev Neurosci.* 2009 January; 10 (1): pp. 58–70.

Enck, P. et al.: 'Therapy Options in Irritable Bowel Syndrome'. In: *Eur J Gastroenterol Hepatol.* 2010 December; 22 (12): pp. 1402–11.

Furness, J.B. et al: 'The Intestine as a Sensory Organ: Neural, Endocrine, and Immune Responses'. In: *Am J Physiol*

Gastrointest Liver Physiol. 1999; 227 (5) pp. G922–G928.

Huerta-Franco, M.R. et al.: 'Effect of Psychological Stress on Gastric Motility Assessed by Electrical Bio-Impedance'. In: *World J Gastroenterol.* 2012 September 28; 18 (36): pp. 5027–33.

Kell, C.A. et al.: 'The Sensory Cortical Representation of the Human Penis: Revisiting Somatotopy in the Male Homunculus'. In: *J Neurosci.* 2005 June 22; 25 (25): pp. 5984–87.

Keller, J. et al.: 'S3-Leitline der Deutschen Gesellschaft für Verdauungs- und Stoffwechselkrankheiten (DGVS) und der Deutschen Gesellschaft für Neurogastroenterologie und Motilität (DGNM) zu Definition, Pathophysiologie, Diagnostik und Therapie intestinaler Motilitätsstörungen' [*S3 Guidelines of the German Society for Digestive and Metabolic Diseases (DGVS) and the German Society for Neurogastroenterology and Motility (DGNM) for the Definition, Pathophysiology, Diagnosis and Treatment of Intestinal Motility Disorders*]. In: *Z Gastroenterol.* 2011; 49: pp. 374–90. (In German)

Keywood, C. et al.: 'A Proof of Concept Study Evaluating the Effect of ADX10059, a Metabotropic Glutamate Receptor-5 Negative Allosteric Modulator, on Acid Exposure and Symptoms in Gastro-Oesophageal Reflux Disease'. In: *Gut.* 2009 September; 58 (9): pp. 1192–99.

Krammer, H. et al.: 'Tabuthema Obstipation: Welche Rolle spielen Lebensgewohnheiten, Ernährung, Prä- und Probiotika sowie Laxanzien?' [*Taboo Subject Constipation: What Part Do Habits, Diet, Pre- and Probiotics and Laxatives Play?*]. In: *Aktuelle Ernährungsmedizin.* 2009; 34 (1): pp. 38–46. (In German)

Layer, P. et al.: 'S3-Leitlinie Reizdarmsyndrom: Definition, Pathophysiologie, Diagnostik und Therapie. Gemeinsame Leitlinie der Deutschen Gesellschaft für Verdauungs- und Stoffwechselkrankheiten (DGVS) und der Deutschen Gesellschaft für Neurogastroenterologie und Motilität (DGNM)' [*S3 Guideline on Irritable Bowel Syndrom: Joint Guideline of the*

German Society for Digestive and Metabolic Diseases (DGVS) and the German Society for Neurogastroenterology and Motility (DGNM)]. In: *Z Gastroenterol.* 2011; 49: pp. 237–93.

Ma, X. et al.: 'Lactobacillus Reuteri Ingestion Prevents Hyperexcitability of Colonic DRG Neurons Induced by Noxious Stimuli'. In: *Am J Physiol Gastrointest Liver Physiol.* 2009 April; 296 (4): pp. G868–G875.

Mayer, E.A.: 'Gut Feelings: The Emerging Biology of Gut-Brain Communication'. In: *Nat Rev Neurosci.* 2011 July 13; 12 (8): pp. 453–66.

Mayer, E.A. et al.: 'Brain Imaging Approaches to the Study of Functional GI Disorders: A Rome Working Team Report'. In: *Neurogastroenterol Motil.* 2009 June; 21 (6): pp. 579–96.

Moser, G. (Ed.): *Psychosomatik in der Gastroenterologie und Hepatologie* [Psychosomatics in Gastroenterology and Hepatology]. Vienna; New York: Springer, 2007. (In German)

Naliboff, B.D. et al.: 'Evidence for Two Distinct Perceptual Alterations in Irritable Bowel Syndrom'. In: *Gut.* 1997 October; 41 (4): pp. 505–12.

Palatty, P.L. et al.: 'Ginger in the Prevention of Nausea and Vomiting: A Review'. In: *Crit Rev Food Sci Nutr.* 2013; 53 (7): pp. 659–69.

Reveiller, M. et al.: 'Bile Exposure Inhibits Expression of Squamous Differentiation Genes in Human Esophageal Epithelial Cells'. In: *Ann Surg.* 2012 June; 255 (6): pp. 1113–20.

Revenstorf, D.: *Expertise zur wissenschaftlichen Evidenz der Hypnotherapie* [Expert Opinion on the Scientific Evidence for Hypnotherapy]. Tübingen, 2003; at http://www.meg.tuebingen.de/downloads/Expertise.pdf (Retrieved October 2013) (In German)

Simons, C.C. et al.: 'Bowel Movement and Constipation Frequencies and the Risk of Colorectal Cancer among Men in the Netherlands Cohort Study on Diet and Cancer'. In: *Am J Epidemiol.* 2010 December 15; 172 (12): pp. 1404–14.

Streitberger, K. et al.: 'Acupuncture Compared to Placebo-Acupuncture for Postoperative Nausea and Vomiting Prophylaxis: A Randomised Placebo-Controlled Patient and Observer Blind Trial'. In: *Anaesthesia*. 2004 February; 59 (2): pp. 142–49.

Tillisch, K. et al.: 'Consumption of Fermented Milk Product with Probiotic Modulates Brain Activity'. In: *Gastroenterology*. 2013 June; 144 (7): pp. 1394–1401.

Part Three

Aggarwal, J. et al.: 'Probiotics and their Effects on Metabolic Diseases: An Update'. In: *J Clin Diagn Res*. 2013 January; 7 (1): pp. 173–177.

Akkasheh, G. et al.: 'Clinical and Metabolic Response to Probiotic Administration in Patients with Major Depressive Disorder: A Randomized, Double-Blind, Placebo-Controlled Trial'. 2016. In: Nutrition 32: pp. 315–320.

Allen, A. P. et al.: 'Bifidobacterium longum 1714 as a Translational Psychobiotic: Modulation of Stress, Electrophysiology and Neurocognition in Healthy Volunteers'. Transl Psychiatry, 2016. 6, e939; doi:10.1038/tp.2016.191.

Arnold, I.C. et al.: '*Helicobacter pylori* Infection Prevents Allergic Asthma in Mouse Models through the Induction of Regulatory T Cells'. In: *J Clin Invest*. 2011 August; 121 (8): pp. 3088–93.

Arumugam, M. et al.: 'Enterotypes of the Human Gut Microbiome'. In: *Nature*. 2011 May 12; 474 (7353); 1: pp. 174–80.

Bäckhed, F.: 'Addressing the Gut Microbiome and Implications for Obesity'. In: *International Dairy Journal*. 2012; 20 (4): pp. 259–61.

Balakrishnan, M.; Floch, M.H.: 'Prebiotics, Probiotics and Digestive Health'. In: *Curr Opin Clin Nutr Metab Care*. 2012 November; 15 (6): pp. 580–85.

Barros, F.C.: 'Cesarean Section and Risk of Obesity in Childhood, Adolescence, and Early Adulthood: Evidence from 3 Brazilian Birth Cohorts'. In: *Am J Clin Nutr.* 2012; 95 (2): pp. 465–70.

Bartolomeo, F. Di: 'Prebiotics to Fight Diseases: Reality of Fiction?'. In: *Phytother Res.* 2013 October; 27 (10): pp. 1457–73.

Benton, D. et al.: 'Impact of Consuming a Milk Drink Containing a Probiotic on Mood and Cognition'. Eur J Clin Nutr 2007; 61: pp. 355–361.

Bischoff, S.C.; Köchling, K.: 'Pro- und Präbiotika' [*Pro- and Prebiotics*]. In: *Zeitschrift für Stoffwechselforschung, klinische Ernährung und Diätik.* 2012; 37: pp. 287–304.

Borody, T.J. et al.: 'Fecal Microbiota Transplantation: Indications, Methods, Evidence, and Future Directions'. In: *Curr Gastroenterol Rep.* 2013; 15 (8): p. 337.

Bräunig, J.: *Verbrauchertipps zu Lebensmittelhygiene, Reinigung und Desinfektion* [Consumer Tips on Food Hygiene, Cleaning and Disinfection]. Berlin: German Federal Institute for Risk Assessment, 2005. (In German)

Brede C.: *Das Instrument der Sauberkeit. Die Entwicklung der Massenproduktion von Feinseifen in Deutschland 1850 to 2000* [The Instrument of Cleanliness. The Development of Fine Soap Mass Production in Germany 1850 to 2000]. Münster et al.: Waxmann, 2005. (In German)

Caporaso, J.G. et al.: 'Moving Pictures of the Human Microbiome'. In: *Genome Biol.* 2011; 12 (5): p. R50.

Carvalho, B.M.; Saad, M.J.: 'Influence of Gut Microbiota on Subclinical Inflammation and Insulin Resistance'. In: *Mediators Inflamm.* 2013; 2013: 986734.

Charalampopoulos, D.; Rastall, R.A.: 'Prebiotics in Foods'. In: *Current Opinion on Biotechnology.* 2012; 23 (2): pp. 187–91.

Chen, Y. et al.: 'Association between *Helicobacter pylori* and Mortality in the NHANES III Study'. In: *Gut.* 2013 September; 62 (9): pp. 1262–69.

Devaraj, S. et al.: 'The Human Gut Microbiome and Body Metabolism: Implications for Obesity and Diabetes'. In: *Clin Chem.* 2013 April; 59 (4): pp. 617–28.

Diop, L. et al.: 'Probiotic Food Supplement Reduces Stress-Induced Gastrointestinal Symptoms in Volunteers: A Double-Blind, Placebo- Controlled, Randomized Trial'. 2008. *Nutrition Research* 28: 1. 1.

Dominguez-Bello, M.G. et al.: 'Development of the Human Gastrointestinal Microbiota and Insights from High-throughput Sequencing'. In: *Gastroenterology.* 2011 May; 140 (6): pp. 1713–91.

Douglas, L.C.; Sanders, M.E.: 'Probiotics and Prebiotics in Dietetics Practice'. In: *J Am Diet Assoc.* 2008 March; 108 (3): pp. 510–21.

Eppinger, M. Et al.: 'Who Ate Whom? Adaptive *Helicobacter* Genomic Changes That Accompanied a Host Jump from Early Humans to Large Felines'. In: *PLoS Genet.* 2006 July; 2 (7): p. e120.

Fahey, J.W. et al.: 'Urease from *Helicobacter pylori* Is Inactivated by Sulforaphane and Other Isothiocyanates'. In: *Biochem Biophys Res Commun.* 2013 May 24; 435 (1): pp. 1–7.

Flegr, J.: 'Influence of Latent Toxoplasma Infection on Human Personality, Physiology and Morphology: Pros and Cons of the Toxoplasma-Human Model in Studying the Manipulation Hypothesis'. In: *J Exp Biol.* 2013 January 1; 216 (Pt. 1): pp. 127–33.

Flegr, J. et al.: 'Increased Incidence of Traffic Accidents in Toxoplasma-Infected Military Drivers and Protective Effect RhD Molecule Revealed by a Large-Scale Prospective Cohort Study'. In: *BMC Infect Dis.* 2009 May 26; 9: p. 72.

Flint, H.J.: 'Obesity and the Gut Microbiota'. In: *J Clin Gastroenterol.* 2011 November; 45 (Suppl.): pp. 128–32.

Fouhy, F. Et al.: 'High-Throughput Sequencing Reveals the Incomplete, Short-Term Recovery of Infant Gut Microbiota following Parenteral Antibiotic Treatment with Ampicillin and Gentamicin'. In: *Antimicrob Agents Chemother.* 2012 November; 56 (11): pp. 5811–20.

Fuhrer, A. et al.: 'Milk Sialyllactose Influences Colitis in Mice Through Selective Intestinal Bacterial Colonization'. In: *J Exp Med.* 2010 December 20; 207 (13): pp. 2843–2854.

Gale, E.A.M.: 'A Missing Link in the Hygiene Hypothesis?'. In: *Diabetologia.* 2002; 45 (4): pp. 588–94.

Ganal, S.C. et al.: 'Priming of Natural Killer Cells by Non-Mucosal Mononuclear Phagocytes Requires Instructive Signals from the Commensal Microbiota'. In: *Immunity.* 2012 July 27; 37 (1): pp. 171–86.

German Federal Government: 'Antwort der Bundesregierung auf die Kleine Anfrage der Abgeordneten Friedrich Ostendorff, Bärbel Höhn, Nicole Maisch, weiterer Abgeordeter und der Fraktion BÜNDNIS 90/DIE GRÜNEN — Drucksache 17/10017. Daten zur Antibiotikavergabe in Nutztierhaltungen und zum Eintrag von Antibiotika und multiresistenten Keimen in die Umwelt. Drucksache 17/10313 [*The German Federal Government's response to the minor interpellation from members Friedrich Ostendorff, Bärbel Höhn, Nicole Maisch, other members, and the ALLIANCE 90/THE GREENS parliamentary group — Parliamentary record 17/10017. Data on the use of antibiotics in livestock farming and the contamination of the environment by antibiotics and multidrug resistant bacteria. Parliamentary record 17/10313*], 17 July 2012, at http://dip21.bundestag.de/dip21/btd/17/103/1710313.pdf (Retrieved October 2013) (In German)

Gibney, M.J.; Burstyn, P.G.: 'Milk, Serum Cholesterol, and the Maasai — A Hypothesis'. In: *Atherosclerosis.* 1980; 35 (3): pp. 339–43.

Gleeson, M. et al.: 'Daily Probiotic's (Lactobacillus casei Shirota) Reduction of Infection Incidence in Athletes'. In: *Int J Sport Nutr Exerc Metab.* 2011 February; 21 (1): pp. 55–64.

Goldin, B.R.; Gorbach, S.L.: 'Clinical Indications for Probiotics: An Overview'. In: *Clinical Infectious Diseases*. 2008; 46 (Suppl. 2): pp. S96–S100.

Gorkiewicz, G.: 'Contributions of the Physiological Gut Microflora to Health and Disease'. In: *J Gastroenterol Hepatol Erkr*. 2009; 7 (1): pp. 15–18.

Grewe, K.: *Prävalenz von Salmonella spp. in der primären Geflügelproduktion und Broilerschlachtung — Salmonelleneintrag bei Schlachtgeflügel während des Schlachtprozesses* [Prevalence of Salmonella spp. in Primary Poultry Production and the Slaughter of Broiler Chickens — Salmonella Contamination of Slaughter Poultry during the Slaughtering Process]. Hanover: Hanover University of Veterinary Medicine, 2011.

Guseo. A.: 'The Parkinson Puzzle'. In: *Orv Hetil*. 2012 December 30; 153 (52): pp. 2060–69.

Herbarth, O. et al.: '*Helicobacter pylori* Colonisation and Eczema'. In: *Journal of Epidemiology and Community Health*. 2007; 61 (7): pp. 638–40.

Hullar, M.A.; Lampe, J.W.: 'The Gut Microbiome and Obesity'. In: *Nestle Nutr Inst Workshop Ser*. 2012; 73: pp. 67–79.

Jernberg, C. et al.: 'Long-Term Impacts of Antibiotic Exposure on the Human Intestinal Microbiota'. In: *Microbiology*. 2010 November; 156 (Pt. 11): pp. 3216–23.

Jin, C.; Flavell, R.A.: 'Innate Sensors of Pathogen and Stress: Linking Inflammation to Obesity'. In: *J Allergy Clin Immunol*. 2013 August; 132 (2): pp. 287–94.

Jirillo, E. et al.: 'Heathy Effects Exerted by Prebiotics, Probiotics, and Symbiotics with Special Reference to Their Impact on the Immune System'. In: *Int J Vitam Nutr Res*. 2012 June; 82 (3): pp. 200–08.

Jones, M.L. et al.: 'Cholesterol-Lowering Efficacy of a Micro encapsulated Bile Salt Hydrolase-Active Lactobacillus reuteri NIMB 30242 Yoghurt Formulation in Hypercholesterolaemic Adults'. In: *British Journal of Nutrition*. 2012; 107 (10) pp. 1505–13.

Jumpertz, R. et al.: 'Energy-Balance Studies Reveal Associations between Gut Microbes, Caloric Load, and Nutrient Absorption in Humans'. In: *Am J Clin Nutr.* 2011; 94 (1): pp. 58–65.

Kato-Kataoka, A. et al.: 'Fermented Milk Containing *Lactobacillus casei* Strain Shirota Preserves the Diversity of the Gut Microbiota and Relieves Abdominal Dysfunction in Healthy Medical Students Exposed to Academic Stress'. 2016. Appl Environ Microbiol 82: pp. 3649–3658.

Katz, S.E.: *The Art of Fermentation: An In-Depth Exploration of Essential Concepts and Processes from around the World.* Chelsea: Chelsea Green Publishing, 2012.

Kelly, J. R. et al.: 'Transferring the Blues: Depression-Associated Gut Microbiota Induces Neurobehavioural Changes in the Rat'. 2016. In: *Psychiatr. Res.* 82, pp. 109–118.

Kountouras, J. et al.: '*Helicobacter pylori* Infection and Parkinson's Disease: Apoptosis as an Underlying Common Contributor'. In: *Eur J Neurol.* 2012 June; 19 (6): p. e56.

Kruijt, A. W. et al.: 'Cognitive Reactivity, Implicit Associations, and theIncidence of Depression: a Two-Year Prospective Study'. 2013. PL oS One 8 (7), e70245.

Krznarica, Z. et al.: 'Gut Microbiota and Obesity'. In: *Dig Dis.* 2012; 30: p. 196–200.

Kumar, M. et al.: 'Cholesterol-Lowering Probiotics as Potential Biotherapeutics for Metabolic Diseases'. In: *Exp Diabetes Res.* 2012; 2012: 902917.

Macfarlane, G.T. et al.: 'Bacterial Metabolism and Health-Related Effects of Galactooligosaccharides and Other Prebiotics'. In: *J Appl Microbiol.* 2008 February; 104 (2): pp. 305–44.

Mann, G.V.: et al.: 'Atherosclerosis in the Masai'. In: *American Journal of Epidemiology.* 1972; 95 (1): pp. 26–37.

Marshall, B.J.: 'Unidentified Curved Bacillus on Gastric Epithelium in Active Chronic Gastritis'. In: *Lancet.* 1983 June 4; 1 (8336) pp. 1273 ff.

Martinson, V.G. et al.: 'A Simple and Distinctive Microbiota Associated with Honey Bees and Bumble Bees'. In: *Mol Ecol.* 2011 February; 20 (3): pp. 619–28.

Matamoros, S. et al.: 'Development of Intestinal Microbiota in Infants and its Impact on Health'. In: *Trends Microbiol.* 2013 April; 21 (4): pp. 167–73.

McKean, J. et al.: 'Probiotics and Subclinical Psychological Symptoms in Healthy Participants: A Systematic Review and Meta-Analysis'. J. Altern Complement Med. 2016; November 2014.

Messaoudi, M. et al.: 'Beneficial Psychological Effects of a Probiotic Formulation (Lactobacillus helveticus R0052 and Bifidobacterium longum R0175)'. In: *Healthy Human Volunteers.* Gut Microbes 2011; 2: pp. 256–261.

Moodley, Y. et al.: 'The Peopling of the Pacific from a Bacterial Perspective'. In: *Science.* 2009 January 23; 323 (5913): pp. 527–30.

Mori, K. et al.: 'Does the Gut Microbiota Trigger Hashimoto's Thyroiditis?'. In: *Discov Med.* 2012 November; 14 (78): pp. 321–26.

Musso, G. et al.: 'Gut Microbiota as a Regulator of Energy Homeostasis and Ectopic Fat Deposition: Mechanisms and Implications for Metabolic Disorders'. In: *Current Opinion in Lipidology.* 2010; 21 (1): pp. 76–83.

Nagpal, R. et al.: 'Probiotics, their Health Benefits and Applications for Developing Healthier Foods: A Review'. In: *FEMS Microbiol Lett.* 2012 September; 334 (1): pp. 1–15.

Nakamura, Y.K.; Omaye, S.T.: 'Metabolic Diseases and Pro- and Prebiotics: Mechanistic Insights'. In: *Nutr Metab (Lond).* 2012 June 19; 9 (1): p. 60.

Nicola, J.P. et al.: 'Functional Toll-Like Receptor 4 Conferring Lipopolysaccharide Responsiveness is Expressed in Thyroid Cells'. In: *Endocrinology.* 2009 January; 150 (1): pp. 500–08.

Nielsen, H.H. et al.: 'Treatment for *Helicobacter pylori* Infection and Risk of Parkinson's Disease in Denmark'. In: *Eur J Neurol.* 2012 June; 19 (6): pp. 864–69.

Norris, V. et al.: 'Bacteria Control Host Appetites'. In: *J Bacteriol.* 2013 February; 195 (3): pp. 411–16.

Okusaga, O.; Postolache, T.T.: '*Toxoplasma gondii*, the Immune System, and Suicidal Behavior'. In: Dwivedi. Y. (Ed.): *The Neurological Basis of Suicide.* Boca Raton, Florida: CRC Press, 2012: pp. 159–94.

Ottman, N. et al.: 'The Function of our Microbiota: Who Is Out There and What Do They Do?'. In: *Front Cell Infect Microbiol.* 2012 August 9; 2: p. 104.

Pavlovíc, N. et al.: 'Probiotics-Interactions with Bile Acids and Impact on Cholesterol Metabolism'. In: *Appl Biochem Biotechnol.* 2012; 168: pp. 1880–95.

Petrof E.O. et al.: 'Stool Substitute Transplant Therapy for the Eradications of *Clostridium difficile* Infection 'RePOOPulating' the Gut'. In: *Microbiome.* 2013 January 9; 1 (1): p. 3.

Reading, N.C.; Kasper, D.L.: 'The Starting Lineup: Key Microbial Players in Intestinal Immunity and Homeostasis'. In: *Front Microbiol.* 2011 July 7; 2: p. 148.

Roberfroid, M. et al.: 'Prebiotic Effects: Metabolic and Health Benefits'. In: *Br J Nutr.* 2010 August; 104 (Suppl. 2): pp. S1–S63.

Romijn, A. R. et al.: 'Double-Blind, Randomized, Placebo-Controlled Trial of Lactobacillus helveticus and Bifidobacterium longum for the Symptoms of Depression'. 2017. In: *Australian & New Zealand Journal of Psychiatry*, pp. 1–12.

Sanders, M.E. et al.: 'An Update on the Use and Investigation of Probiotics in Health and Disease'. In: *Gut.* 2013; 62 (5): pp. 787–96.

Sanza, Y. et al.: 'Understanding the Role of Gut Microbes and Probiotics in Obesity: How Far Are We?'. In: *Pharmacol Res.* 2013 March; 69 (1): pp. 144–55.

Sarkar, A. et al.: 'Psychobiotics and the Manipulation of Bacteria-Gut- Brain Signals'. Trends Neurosci. November 2016; 39(11): pp. 763–781.

Schmidt, C.: 'The Startup Bugs'. In: *Nat Biotechnol.* 2013 April; 31 (4): pp. 279–81.

Scholz-Ahrens, K. E. et al.: 'Prebiotics, Probiotics, and Synbiotics Affect Mineral Absorption, Bone Mineral Content, and Bone Structure'. In: *J Nutr.* 2007 March 137 (3 Suppl. 2): pp. 838S–846S.

Schwarz, S. et al.: 'Horizontal versus Familial Transmission of *Helicobacter pylori*'. In: *PLoS Pathog.* 2008 October; 4 (10): p. e1000180.

Shen, J. et al.: 'The Gut Microbiota, Obesity and Insulin Resistance'. In: *Mol Aspects Med.* 2013 February; 34 (1): pp. 39–58.

Starkenmann, C. et al.: 'Olfactory Perception of Cysteine-S-Conjugates from Fruits and Vegetables'. In: *J Agric Food Chem.* 2008 October 22; 56 (20): pp. 9575–80.

Steenbergen, L. et al.: 'A Randomized Controlled Trial to Test the Effect of Multispecies Probiotics on Cognitive Reactivity to Sad Mood'. 2015. *Brain, Behavior, and Immunity* 48: pp. 258–264.

Stowell, S.R.: 'Innate Immune Lectins Kill Bacteria Expressing Blood Group Antigen'. In: *Nat Med.* 2010 March; 16 (3): pp. 295–301.

Tängdén, T. et al.: 'Foreign Travel is a Major Risk Factor for Colonization with Escherichia coli Producing CTX-M-Type Extended-Spectrum ß-Lactamases: A Prospective Study with Swedish Volunteers'. In: *Antimicrob Agents Chemother.* 2010 September; 54 (9): pp. 3564–68.

Teixeira, T.F.: 'Potential Mechanisms for the Emerging Link between Obesity and Increased Intestinal Permeability'. In: *Nutr Res.* 2012 September; 32 (9): pp. 637–47.

Torrey, E.F. et al.: 'Antibodies to *Toxoplasma gondii* in Patients with Schizophrenia: A Meta-Analysis'. In: *Schizophr Bull.* 2007 May; 33 (3): pp. 729–36.

Tremaroli, V.; Bäckhed, F.: 'Functional Interactions between the Gut Microbiota and Host Metabolism'. In: *Nature*. 2012 September 13; 489 (7415): pp. 242–49.

Turnbaugh, P.J.; Gordon, J.L.: 'The Core Gut Microbiome, Energy Balance and Obesity'. In: *J Physiol*. 2009; 587 (17): pp. 4153–58.

de Vrese, M.; Schrezenmeir, J.: 'Probiotics, Prebiotics, and Synbtiotics'. In: *Adv Biochem Engin/Biotechnol*. 2008; 111: pp. 1–66.

de Vriese, J.: 'Medical Research. The Promise of Poop'. In: *Science*. 2013 August 30; 341 (6149): pp. 954–57.

Vyas, U.; Ranganathan, N.: 'Probiotics, Prebiotics and Synbiotics: Gut and Beyond'. In: *Gastroenterol Res Pract*. 2012; 2012: 872716.

Webster, J.P. et al.: 'Effect of *Toxoplasma gondii* upon Neophobic Behaviour in Wild Brown Rats, Rattus norvegicus'. In: *Parasitology*. 1994 July; 109 (pt. 1): pp. 37–43.

Wichmann-Schauer, H.: *Verbrauchertipps: Schutz vor Lebensmittelinfektionen im Privathaushalt* [Tips: Protection against Food Infections in Private Households]. Berlin: German Federal Institute for Risk Assessment, 2007. (In German)

Wu, G.D. et al.: 'Linking Long-Term Dietary Patterns with Gut Microbial Enterotypes'. In: *Science*. 2011 October 7; 334 (6052): pp. 105–08.

Yatsunenko, T. et al.: 'Human Gut Microbiome Viewed Across Age and Geography'. In: *Nature*. 2012 May 9; 486 (7402): pp. 222–27.

Zipris, D.: 'The Interplay between the Gut Microbiota and the Immune System in the Mechanism of Type 1 Diabetes'. In: *Curr Opin Endocrinol Diabetes Obes*. 2013 August; 20 (4): pp. 265–70.